AN APPLE A DAY

AN APPLE A DAY

THE MYTHS, MISCONCEPTIONS, AND TRUTHS ABOUT THE FOODS WE EAT

Joe Schwarcz, Ph.D.

OTHER PRESS
NEW YORK

Production Editor: Yvonne E. Cárdenas

Book design: Simon M. Sullivan

This book was set in 11.25 pt Berkeley
by Alpha Design & Composition of Pittsfield, NH.

10 9 8 7 6 5 4 3 2 1

Library of Congress Cataloging-in-Publication Data

Schwarcz, Joseph A.
An apple a day : the myths, misconceptions, and truths about
the foods we eat / Joe Schwarcz.
p. cm.
Originally published: Toronto : HarperCollins, 2007.
Includes index.
ISBN 978-1-59051-311-8
1. Nutrition—Popular works.   2. Food—Popular works.
3. Food additives—Popular works.   I. Title.
TX355.S3475 2009
363.8—dc22   2008021638

It is a capital mistake to theorize before one has the data. Insensibly one begins to twist facts to suit theories, instead of theories to suit facts.

—SHERLOCK HOLMES

Almost all aspects of life are engineered at the molecular level, and without understanding molecules we can only have a very sketchy understanding of life itself.

—FRANCIS CRICK

CONTENTS

PART TWO
Manipulating Our Food Supply

PART THREE
Contaminants in Our Food Supply

PART FOUR
Tough to Swallow

AN APPLE A DAY

INTRODUCTION

Eating used to be simple. As long as the food was tasty, looked reasonably appetizing and was plentiful, we were content. But then science came to dinner, and all of a sudden sitting down at the table became a laboratory experience, and a confusing one at that. Eat fish, we were told, it's chock full of omega-3 fats. Be careful, urged another report, fish may harbor "good" fats, but it's also loaded with PCBs and mercury. We switched to margarine from butter because it had fewer saturated fats. But then came accusations that the trans-fatty acids it contained clogged arteries just like saturated fats. Eat soy, we were told, it lowers your cholesterol. Don't eat soy, it affects thyroid function. Drink milk, you need the calcium. Don't drink milk, it forms mucus. Drink coffee, it is full of antioxidants. Don't drink coffee, it raises blood pressure. Then there are those little gems that "they" say. Stay away from MSG. Don't touch foods preserved with nitrites. Or with sulphites. Beware of pesticide residues. Ban foods that contain genetically modified organisms. Don't cook in Teflon pots. Or in microwave ovens. Stay away from sugar. And don't even think about artificial sweeteners! But just who are "they"? We've been advised to load up on foods like oats, flax, mangosteen juice, garlic, and oregano because some researchers showed each of these to be of some benefit. Whole-grain bread may be in one day—lots of useful fiber and vitamins—but out the next because acrylamide, a purported carcinogen, was detected in the crust.

Many people throw up their arms in bewilderment at all this confusing nutritional information, and go back to their old dietary regimens. And that is too bad. Nutrition *is* important. The challenge is to separate the wheat from the chaff, and come to some practical conclusions about what to eat, based not on hearsay but on sound science.

This is not easy to do, especially when we consider that the human body is the most complex machine on the face of the earth. The diversity of its molecular components is so astounding that by comparison, computers, medical scanners, and space vehicles are simple devices. What we call life is really the result of astonishingly complex molecular activity that goes on in every cell in our body every second of the day. And where do the molecules that engage in these intricate gymnastics come from? One way or another, they come from the food we eat.

It seems obvious then that the composition of our diet can affect our molecular makeup, and consequently our health. But the relationship between diet and health is not a simple one. Food is chemically very complex. An apple, for example, is composed of over 300 different compounds. A single meal can flood the body with thousands of compounds, many of which have never been isolated or identified. While nutrition is obviously an important determinant of health, it is folly to think that one can introduce something as complicated as food into something as complex as the human body and make easy predictions about the outcome. Curing disease by dietary manipulation must therefore be looked at with a healthy degree of skepticism. But *preventing* disease by modifying our diet is realistic. The question is: how?

Separating sense from nonsense has been the focus of my educational activities ever since I began teaching chemistry back in 1973. This book does not purport to be an encyclopedia of nutrition or a comprehensive guide to healthy eating. It does, however, provide a framework for sound nutritional thinking, along with a perspective on what is worth worrying about and what is not as we ply ourselves with the mélange of molecules we call food.

People have different appetites when it comes to sampling food issues. Some are interested in the nutritional merits of specific foods; some are fascinated by the action of antioxidants; and others worry about the safety of additives. Chances are that most of you will be as picky with this book as you are with your food. Each of the chapters in this volume has been designed to stand on its own and to provide you with up-to-date information on a specific food-related issue. In Part One, we examine the role played by food's natural components. What is there in tomatoes, soy, or broccoli that can contribute to good health? Why does gluten in wheat cause problems for some people? Part Two investigates the consequences of human intervention in our food supply. What are the risks and benefits of food additives or of genetic modification? What promise lies in adding specific bacteria to foods? Part Three deals with substances—such as pesticide residues, remnants of antibiotics, trans fats, and chemicals from plastics—that end up in our food supply unintentionally as a result of food processing. And for good measure, after you've waded through the science, in Part Four, we'll throw in a discussion of some dubious nutritional ideas.

These are all fascinating issues. Now, on to the fun part: let's try to digest them.

PART ONE

NATURALLY OCCURRING SUBSTANCES
IN OUR FOOD SUPPLY

AN APPLE A DAY

Is there a better subject with which to begin a discussion of the relationship between food and health than apples? After all, doesn't "an apple a day keep the doctor away"? Maybe it does, if you throw it at her! There is no single food that has magical health properties. There are good diets and there are bad diets. It is certainly possible to have a good diet and never eat apples, just as it is possible to gorge on apples and have a horrible diet. What really matters in terms of nutrition is the net effect produced by all of the chemicals that wend their way into our bodies from the food we eat. Yes, chemicals. I can practically see those eyebrows being raised. It may seem unusual to see the word *chemical* without an adjective like *poisonous*, in front of it. Actually, without appropriate context, *toxic chemical* is a meaningless term.

Take salicylic acid as an example. It occurs naturally in a variety of fruits and plants, including apples. It is also formed in our body when aspirin is metabolized. Indeed, salicylic acid is responsible for the physiological effects of aspirin, which include reducing the risk of blood-clot formation. That's why aspirin is used to treat a heart attack, and why it is commonly taken in small doses to prevent one. But in an overdose, salicylic acid can kill. Before childproof packaging was introduced, aspirin poisoning was a common cause of death in children. So how do we react if a test detects salicylic acid in our blood? Panic because of the presence of a "toxic chemical," or relief because of possible protection against heart disease? Of course, without the proper context there can be no appropriate reaction. To decide whether to laugh or cry, we need to know what blood levels of salicylic acid have been linked to risk and what levels to protection from disease. The mere presence of the chemical says nothing. As Paracelsus insightfully and wisely noted some 500 years ago, "Only the dose makes the poison." And to this we can add, "And only the dose makes the cure!"

So let's not get paranoid about chemicals in our food. Everything in the world is made of chemicals, and if you restricted yourself to a diet free of chemicals, you would be dining in a vacuum! With that in mind, let's investigate the chemicals in an apple. So tell me, would you like some nail polish remover in your diet? Or rubbing alcohol? Then have an apple! Yes, all apples contain acetone and isopropanol. And if these don't sound toxic enough, you can throw in some cyanide. It's there too. Added by nature, not by humans! Should you then be worried about eating apples? Of course not! The amounts of these chemicals are too small to be of any consequence. Apples, as already mentioned, contain over 300 naturally occurring compounds, and whatever effect the fruit has on our health is a reflection of all of these. Researchers are particularly excited about one class of compounds, the polyphenols. Why? Because they have powerful antioxidant properties.

Chances are that if you haven't heard chapter and verse about antioxidants in recent years, you've been spending too much time in the butcher shop. These highly publicized substances are found in fruits and vegetables and can neutralize free radicals, those rogue molecular fragments produced whenever we inhale oxygen. We can't live without oxygen, of course, but there is a cost to be paid for living with it: illness and eventual death! About 2 to 3 percent of the oxygen consumed by our cells is converted into free radicals that are so reactive, they can rip other molecules apart. When the victims are proteins, fats, nucleic acids, or other essential biomolecules, the result can be heart disease, cancer, or dementia. Even plain old aging has been linked to cumulative free-radical damage.

Since antioxidants can mop up excess free radicals, they obviously merit serious scientific investigation. One of the difficulties, though, is the large variety of antioxidants that are present in plant products. Vitamins C and E, along with carotenoids, have received a great deal of attention, but most of the antioxidant activity of fruits and vegetables can be attributed to polyphenols. The term *polyphenol* actually refers to several related families of molecules that include the flavonoids, anthocyanins, chalcones, and hydroxycinnamates. To complicate things further, each family in turn comprises many compounds that are linked by some common feature of their molecular structure. As one might expect, because these antioxidants have different molecular structures, they also have different degrees of antioxidant activity. Obviously, knowledge about the distribution of polyphenols in our diet, coupled with knowledge about which ones have the most activity, would be very useful.

But before we start jumping on the polyphenol bandwagon, we need to ask a pertinent question: What evidence do we have that polyphenols in the diet can contribute to good health? Demonstrating that these chemicals can neutralize free radicals in a test tube is one thing, showing that they can prevent cancer or heart disease is quite another. The first major study to suggest such a possible benefit appeared in *The Lancet* in 1993. Dutch researchers measured the amount of flavonoids in various foods, and by means of a dietary questionnaire assessed the flavonoid intake of 805 men ages sixty-five to eighty-four who were then followed for five years. Even when adjustments were made for smoking, body weight, cholesterol levels, blood pressure, physical activity, and vitamin and fiber intake, the polyphenol content of the diet was inversely associated with death from heart disease. The major sources of polyphenols in this study were tea, onions, and apples. A single apple a day made a difference!

There is evidence for the anticancer effects of polyphenols as well. Researchers at Cornell University found that treating colon or liver cancer cells in the laboratory with apple extract inhibited their proliferation, with extracts from the skin performing even better than extracts from the flesh. The same Cornell team also showed that apples may play a role in reducing the risk of breast cancer. Rats exposed to a substance known to trigger breast cancer were fed apple extract in amounts equivalent to a human eating one, three, or six apples a day. Lo and behold, the chance of developing the disease was reduced by 17, 39, and 44 percent respectively! Even when cancer set in, maintaining the apple diet blocked the spread of the disease, and after six months reduced the number of tumors by 25 percent. And that with just one apple a day! These researchers did not stop at investigating cancer. When they exposed rat brain cells to a specific polyphenol, quercetin, they found that the cells resisted oxidative damage more, implying a potential reduction in the risk of developing Alzheimer's and other such brain diseases. Indeed, a group at the University of South Florida found a greatly reduced risk of Alzheimer's disease in seniors who drank fruit or vegetable juices at least three times a week compared with those who drank these juices less than once a week.

Other studies have found that quercetin reduces the growth of human prostate cancer cells in the lab and that its presence in the diet is inversely associated with the risk of lung cancer. This is not that surprising, given that quercetin has very potent antioxidant activity. And it is found in apples, along, of course, with many other polyphenols. But before we start attributing magical properties to apples, let's realize that there are foods with higher antioxidant potential. Red kidney beans, blueberries, and cranberries all have greater antioxidant capacity per serving. And oregano has forty times the antioxidant activity of apples. What matters, though, is the total intake of polyphenols. Let's face it, eating apples every day is easy. Kidney beans are more challenging.

But the real key to antioxidant intake is variety. The more varied the fruits and vegetables consumed, the greater the chance that we equip ourselves with the complex array of antioxidants that may be needed for good health. Studies indicate we should be aiming for a daily polyphenol intake of around one gram. Apples, depending on the variety, can contribute anywhere from 100 to 300 milligrams. Eating a couple a day is certainly a good idea. And if someone tries to scare you by pointing out that apples contain embalming fluid, you can respond that whatever the detriments of the traces of the naturally occurring formaldehyde may be, they are more than countered by the benefits of the polyphenols. Eat those apples, and make the undertaker wait longer with his embalming fluid.

TOMATOES AND LYCOPENE

Researchers are really excited about lycopene, the compound responsible for the red color of tomatoes. So is the public. Prompted by ads in magazines and by seductive promotions in health food stores, lycopene supplements are enjoying brisk sales, especially among men worried about prostate cancer.

Why should lycopene have any effect on prostate cancer? Because studies have shown that men who consume lots of tomato products have a lower incidence of the disease. A study by the Harvard School of Public Health showed that men who had ten or more servings of tomato-based foods a week had a 45 percent reduction in the rate of prostate cancer. Spaghetti sauce was the most common tomato-based food consumed, and cooked tomatoes seemed to be more protective than raw tomatoes or tomato juice, perhaps because heat releases lycopene and other nutrients from tomato cells. Also, the sauce is commonly made with olive oil, which enhances the absorption of the fat-soluble lycopene. And the sauce is a concentrated tomato product, so it provides more nutrients per gram than fresh tomatoes.

Lycopene is a good candidate for biological activity because the tomato actually uses the compound to maintain its own health. It protects the seeds in the fruit from damage by oxygen and light. Lycopene can absorb ultraviolet light, and its antioxidant activity allows it to neutralize free radicals generated by exposure to oxygen. Of course, there's more to tomatoes than lycopene. Like other plant products, tomatoes are very complex chemically and contain hundreds of different compounds. Is lycopene the most important one? Researchers at Ohio State University decided to find out.

Since triggering cancer in humans is out of the question, researchers focused on rats, which actually are very good models for human prostate cancer. They caused prostate cancer in about 200 rats by treating them with a cancer-inducing mix of testosterone and N-methyl-N-nitrosourea. Some of the rats were then fed diets that contained whole-tomato powder while others were treated to rat chow fortified with lycopene. The lycopene-fortified rats were actually getting more lycopene than the tomato-powder rats. That's what made the results of the experiment so surprising. The risk of death from prostate cancer was significantly greater in the rats that were fed the pure lycopene extract! This would seem to suggest that there are other components in tomatoes that have a protective effect and that the whole food is beneficial, while isolated components may not be. True, the study was done in rats, but it does send us a message. Eat a balanced diet, with lots of vegetables and fruits, because shortcuts may not work.

There was another significant finding in this study. The researchers also put some of the rats in each group on a calorie-restricted diet. While their companions were allowed to eat as much as they wanted, these rats were given a diet that contained 20 percent fewer calories than what rats usually consume. Guess what? These hungry rats lived longer than the rats who ate freely. So, just eating less food reduces the risk of prostate cancer. What's the overall message for human beings? We should reduce our calorie intake and eat lots of tomato products. And eating those tomato products may even play a role in protecting the heart. At least that's one conclusion that can be drawn from an intriguing Italian study.

Imagine being admitted to a hospital with a heart attack and a doctor asking how many times a week you eat pizza. This was the actual question that was asked of 507 heart attack victims and 478 others who had been admitted to a hospital in Milan, Italy, between

1995 and 1999. Why? To find out if most Italian foods had any role to play in heart disease. We've all heard about the benefits of the highly touted Mediterranean diet, and Italian researchers decided to find out if pizza specifically played a role in protection against cardiovascular disease.

After admission to the hospital, the patients were interviewed about their lifestyle habits and their diets. They filled out a seventy-eight-item food-frequency questionnaire on the basis of which they were divided into non-pizza eaters, occasional pizza eaters (one to four portions a month), and regular pizza eaters (more than one portion per week). Heart attack victims reported they had exercised less than controls, smoked more often, consumed more coffee, and drank less alcohol. No surprise here. They also had more of a history of high blood pressure, consumed more calories, and ate fewer fruits and vegetables. Still no surprise. But the surprise came when pizza eating was considered. Regular pizza eaters were 40 percent less likely to suffer a heart attack than those who never ate pizza! Why this should be so is somewhat of a mystery. Perhaps pizza eating is just an indicator of following a Mediterranean diet, which tends to be lower in fat than the North American diet.

We have to remember that we are talking about pizza as served in Italy, not the American version. No double cheese, no cheese-filled dough, no piles of pepperoni or globs of trans fat–laden shortening. The dough is thin, the pizza is dressed with olive oil and cheese, and there is plenty of fresh tomato sauce. The answer to this pizza mystery may lie not in what the people are eating, but rather in what they are *not* eating. Perhaps the pizza is displacing high-fat hamburgers and fries from the diet. Let's note that a portion of pizza in the Italian study was defined as 200 grams, and even the so-called regular eaters averaged only 500 grams (just over a pound) of pizza per week. Indeed, pizza may be displacing higher-calorie foods from the diet. Or maybe it's the yellow stuff around the tomato seeds that matters. This fluid contains flavonoids that have anticlotting properties, and could, at least in theory, reduce the risk of heart attacks.

The producers of Fruitflow certainly think this is the case. This patented tomato-extract product is being added to various drinks with hopes of improving cardiovascular health. In one study, blood "stickiness" was reduced by an average of 70 percent in 220 volunteers who drank a juice containing Fruitflow, with the effect lasting for eighteen hours. Tomato juice itself may provide a similar benefit. And it may be especially helpful on long-distance flights, where a potentially life-threatening condition called deep-vein thrombosis can occur. Sitting in one position, such as in an airline seat, without moving for extended periods increases the chance of blood clots forming in the legs. These clots can travel to the heart or lungs and cause a catastrophe. So loading up on the tomato juice (without the vodka) is a good idea on long flights. Fruitflow is not the only tomato extract under investigation. Israeli researchers found that a supplement sold as Lyc-O-Mato, capsules that contain nutrients equivalent to those found in about four tomatoes (along with some fat to aid absorption), reduced moderately elevated blood pressure significantly. Hmm . . . tomatoes and fat . . . bring on the pizza! And top it with broccoli!

University of Illinois nutrition professor John Erdman fed a diet containing 10 percent dehydrated tomato powder, or 10 percent broccoli powder, or a combination of both, to rats implanted with human prostate-cancer cells. Another group of rats was treated with supplemental lycopene, and yet another group was castrated, a possible treatment for prostate cancer. After twenty-two weeks, Erdman's team found that the tomato-broccoli combination was most effective in reducing the size of tumors. This was an animal study, so it is more meaningful than a test-tube experiment, but more importantly, the dose of broccoli and tomato needed to achieve the reduction in tumors is within the norms of the human diet. Conversion of the amounts fed to the animals to a human dose suggests that a cup and a half of broccoli a day coupled with two and a half cups of fresh tomato, or a cup of tomato sauce, can be effective in reducing the growth of prostate tumors and probably in reducing their occurrence as well. Why a combination of broccoli and tomato works better than the individual foods is not known, but compounds in foods can inhibit cancer in various ways, ranging from stimulating detoxicating enzymes to triggering cell death. Maybe there is a market out there for broccoli-flavored tomato ketchup.

Eating tomatoes may not only make you healthier, it may also make you look better. Lycopene is fat-soluble and concentrates in fatty tissue, such as the fatty layer just underneath the skin. Since the molecule is an efficient absorber of ultraviolet light, it offers some protection against sun-induced skin damage. In conjunction with the BBC television series *The Truth About Food*, two dermatologists in Britain put this notion to a test. They recruited twenty-three women aged twenty to fifty who were willing to bare their bottoms and expose them to ultraviolet light for the sake of science.

Half the volunteers consumed sixteen milligrams of lycopene daily, the amount contained in three teaspoons of tomato paste, along with ten grams of olive oil to help with absorption of the fat-soluble lycopene. The other volunteers got only the olive oil. Otherwise, both groups had identical diets. The results? Less reddening of the skin and less DNA damage in the lycopene group. And if you don't like tomato paste, a glass of tomato juice or a cup of tomato soup will do. But as for fresh tomatoes, well, you would need to eat at least half a dozen to achieve the same effects.

All of these intriguing studies have prompted producers to petition the Food and Drug Administration in the United States to allow health claims on the labels of tomato products. After all, foods containing soy and oats can have labels that claim they reduce cholesterol, and calcium-supplement labels can claim that the supplements reduce the risk of osteoporosis, so why shouldn't tomato-product labels be allowed to make claims about reducing the risk of cancer? The FDA responds that there simply is not enough evidence to support the cancer-reduction claim. But the FDA does agree that there may be some health benefits to eating tomatoes. So it allows the tomato-product labels to include statements such as this: "Very limited and preliminary scientific research suggests that eating one-half to one cup of tomatoes and/or tomato sauce a week may reduce the risk of prostate cancer. The FDA concludes that there is little scientific evidence supporting this claim." Of course, the tomato producers believe that the FDA is too stringent in its requirements and that there is enough evidence about lycopene to warrant a stronger health claim.

Researchers at the National Cancer Institute and The Fred Hutchinson Cancer Research Center share the FDA's skepticism about lycopene. If lycopene does indeed offer protection from cancer, then people who have higher levels of this compound in their blood should have a reduced risk of cancer. But this does not appear to be the case. Researchers followed over 28,000 men between the ages of fifty-five and seventy-four who had no history of prostate cancer. During eight years of follow-up, 1,320 of the men were diagnosed with prostate cancer, but no relationship was found between the blood levels of lycopene and occurrence of the disease.

Of course, this research does not signal the end of the debate. We cannot just dismiss the studies that have shown an association between consuming tomatoes and protection from cancer. Let's remember that tomatoes are chemically complex and contain numerous compounds besides lycopene, compounds that may—either alone or in combination with lycopene—act as anticancer agents. Perhaps the most important point is that the scientific evidence does not support the concept of a "superfood" or a "super" ingredient. Vegetables, fruits, and whole grains are loaded with compounds that have shown a potential for protection from cancer. Loading up on any single food or supplement is not the answer. The key is to eat a variety of foods that contain these beneficial chemicals, including, of course, tomato products.

Lycopene supplements may be useful in the future, but so far, there is no compelling evidence that these are as effective as tomato products.

CRANBERRIES AND PROCYANIDINS

Cranberries and turkeys make a pretty good combo. But so do cranberries and people. Don't worry, I'm not proposing cannibalism, just a scientific evaluation of the possible health benefits of cranberries.

Mention cranberry juice and "urinary tract infection" springs to mind. Most women and many men are familiar with the frequent urination and accompanying burning sensation that signals a bacterial invasion of the urinary tract. Today antibiotics solve the problem, but what did people do before? "Flushing the system" seemed a logical approach. I suppose all sorts of beverages were tried, but by the mid-1800s books on folkloric medicine were suggesting the use of cranberry juice. Based on anecdotal evidence, the juice developed a solid reputation for treating and preventing urinary tract infections.

Once bacteria had been identified as the cause of UTIs, scientists began to explore possible mechanisms by which cranberry juice could offer relief. Acidifying the urine to make it more inhospitable to bacteria was a possibility, as was the antibacterial action of hippuric acid, a component of cranberries. But trying to explain how cranberry juice worked before clearly demonstrating that it did was putting the cart before the horse. Finally, in 1994, Harvard researchers decided to mount a proper clinical study of the claims. They enrolled 153 older women, half of whom were given ten ounces (285 milliliters) of cranberry juice every day, while the other half were given a look-alike drink containing no cranberry. The women who drank cranberry juice were 58 percent less likely to have levels of bacteria in their urine that would be expected to cause infections. As we would eventually learn, the effect was not due to acidity of the urine, nor to the antibacterial effect of hippuric acid. It had to do with compounds that prevented bacteria from adhering to the lining of the urinary tract.

Bacteria produce adhesives that enable them to stick to tissues so they can pick up nutrients more readily. These molecules fit into specific receptor sites on the epithelial cells that line the urinary tract. As was cleverly shown by Yale University researchers in 1994, compounds in cranberries block these receptors. Urine samples were collected from volunteers who were then given four ounces (115 milliliters) of cranberry juice to drink. Four to six hours later urine was again collected and incubated with *E. coli* bacteria, the kind that normally are responsible for urinary tract infections. The experiment was then repeated with eight ounces (230 milliliters) of juice. Separately, the scientists cultured cells taken from the lining of the human bladder and then mixed them with the urine samples. Lo and behold, the bacteria did not stick as effectively to the cells when the urine samples came from women who had consumed cranberry juice! Furthermore, the more juice consumed, the less the bacteria adhered to the cells. Although the specific ingredients in the juice responsible for this effect have not been conclusively identified, speculation is that substances known as trimeric procyanidins may be responsible.

These procyanidins may do more than prevent urinary tract problems. Most ulcers are caused by infection with the *Helicobacter pylori* bacterium. Well, it seems that the procyanidins may also prevent these bacteria from infecting the stomach. Researchers in China chose a population with a high rate of *Helicobacter* infection, and in a placebo-controlled, double-blind study, gave ninety-seven people 500 milliliters (just over two cups) of cranberry juice for ninety days, while ninety-two others got a placebo. They found that *H. pylori* was eradicated in fourteen people in the cranberry group but in only five in the placebo group. Not an earth-shaking difference, but significant nevertheless, especially given the resistance problems we are now encountering with antibiotics.

Cranberries may even help reduce cavities by preventing *Streptococcus mutans* bacteria from clinging to teeth. These are the bacteria that digest sugars and convert them into acids that eat away tooth enamel. When researchers at the University of Rochester used cranberry juice to coat a sample of hydroxyapatite, the material of which tooth enamel is made, they found significant protection against bacterial adhesion. But nobody is suggesting swishing the mouth with cranberry juice. The commercial varieties often have loads of added sugar, not the best thing for the teeth, or indeed for the rest of the body, and you can't drink pure cranberry juice: it's much too sour. That's why researchers are trying to isolate the active ingredients in cranberries and make them available in a capsule form.

There may be even more to cranberries than their effect on bacteria. At the University of Western Ontario, twenty-four female mice were fed a normal diet for twelve weeks, twenty-four others drank cranberry juice instead of water, and another twenty-four had their diet supplemented with cranberry solids (the stuff left after the juice is squeezed) to make up 1 percent of their chow. Then one week later, one million human breast tumor cells were injected into the animals' mammary glands. These were a special breed of mice, genetically engineered to have a compromised immune system, so they all developed cancer. But the ones that drank the juice took two weeks longer to be affected, and the ones that ate the cranberry solids took four weeks longer to develop tumors than the mice on ordinary lab food. Autopsies showed that ingesting cranberry cut—by more than half—the number of tumors that spread to the lungs and lymph nodes. Want even more good news? Researchers at the University of Prince Edward Island have shown that cranberries may even protect against brain damage caused by stroke.

So, you're now reaching for a glass of cranberry juice. Ready for the downside? Cranberries contain compounds that may inhibit enzymes used to break down certain medications. There have been a handful of reports of people on Coumadin (a common blood thinner) developing bleeding disorders after consuming cranberry juice. To be on the safe side, it seems a good idea for anyone on Coumadin to limit the amount of cranberry juice consumed. And one more thing. When researchers studied the ability of scents to stimulate men sexually, they found that the most enticing smell was a combination of lavender and pumpkin pie. The least enticing? Cranberry!

GRAPEFRUIT AND FURANOCOUMARINS

rapefruit growers don't know whether to laugh or cry. There is some evidence that eating the fruit or drinking its juice can reduce blood cholesterol. But there are also those troublesome studies about grapefruit interfering with the effectiveness of certain medications, including some of the statin drugs used to fight high cholesterol. What a conundrum! Do we give up the juice or the drug? As you might expect, the situation is more complicated than it first seems.

"A chance finding of our study on ethanol–drug interactions was that citrus fruit juices may greatly augment the bioavailability of some drugs." So began a paper published in 1991 in *The Lancet*, one of the most respected medical journals in the world. Dr. David Bailey and colleagues at the University of Western Ontario had been studying felodipine, a blood pressure–lowering drug, and wondered if it interacted with alcohol. They decided on a double-blind trial in which some subjects were to take the drug with alcohol and some without. This meant that the taste of alcohol had to be masked, and after some experimentation Dr. Bailey concluded that grapefruit juice was up to the task. To the researchers' surprise, the alcohol had no effect, but in both groups the blood levels of felodipine were three times higher than expected. Bailey knew he was on to something. And he certainly liked a challenge. After all, he had been the first Canadian to run a sub-four-minute mile!

At this point, the eager scientist decided to become his own guinea pig. One day he took felodipine with water, the next day with grapefruit juice; each time he drew blood, and sent it for analysis of drug levels. He didn't have to wait for the results to confirm the "grapefruit effect." After taking the medication with grapefruit juice, he began to feel faint and lightheaded, classic symptoms of low blood pressure. Clearly, when taken with grapefruit juice, felodipine lowered blood pressure more than expected. A number of questions immediately arose. By what mechanism was grapefruit juice increasing the drug's effectiveness? Did other juices have such an effect? What about interactions with other drugs? What would happen if the juice were consumed not with the drug but at some other time during the day? Was there a silver lining to this cloud? Could dosages of medications be reduced if they were taken with grapefruit juice?

Predictably, the *Lancet* paper unleashed a cavalcade of research. It wasn't long before studies showed that only grapefruit juice had this unusual effect. Some compound specific to grapefruit inhibited the action of CYP3A4, an enzyme found in the wall of the intestine. This enzyme is part of the body's detoxicating system and tackles intruders, such as medications. If its action is impaired, blood levels of these foreign substances can be expected to rise. Since CYP3A4 is known to be involved in the metabolism of numerous drugs, researchers suspected that felodipine would not be the sole medication to show a "grapefruit effect." Indeed it was not. Various oral medications, ranging from heart-rhythm regulators and immunosuppressants to estrogen supplements and AIDS treatments, all interact with grapefruit juice. And the effect can last as long as twenty-four hours, meaning that drinking grapefruit juice at any time is contraindicated when taking drugs metabolized by CYP3A4. Since it isn't completely clear which drugs fall into this category and which do not, and because of the known variation in CYP3A4 levels in different individuals, some experts suggest that grapefruit juice be avoided when taking any medication.

Accordingly, many hospitals have taken grapefruit juice off the menu.

The grapefruit industry has complained that it is being unfairly singled out. Spokespeople maintain, correctly, that there are numerous drug–food interactions. Dairy foods can interfere with some antibiotics, broccoli can reduce the effect of anticoagulants, foods high in tyramine (aged cheese, red wine, soy sauce, sauerkraut, salami) can cause dramatic rises in blood pressure when coupled with antidepressants of the monoamine oxidase (MAO)–inhibitor variety, and the absorption of digoxin (taken for congestive heart disease) is impaired by cereals such as oatmeal. While all of this information is factual, the existence of such effects doesn't let grapefruit off the hook.

So if, as mentioned earlier, grapefruit juice lowers cholesterol, why not forget the statin and just drink grapefruit juice? That's exactly what some people are asking after reading about Israeli researcher Shela Gorinstein's study showing that just one red grapefruit a day can reduce LDL, the "bad cholesterol" by as much as 20 percent. Furthermore, red grapefruit reduces triglycerides significantly. But wait a minute. Gorinstein's subjects all had recently undergone bypass surgery, had been resistant to statins, and were on a diet with only 9 percent of calories coming from fat. So these results cannot be readily extrapolated to your average North American who is diagnosed with elevated cholesterol. What can he or she do? First, eat a diet low in saturated and trans fats, with plenty of fruits, vegetables, and oat bran. And yes, grapefruit too! Throw in some persimmons, some pomelits (a cross between a grapefruit and a pomelo), and pale lager beer, all of which have been shown by Gorinstein to reduce the risk of heart disease. And if that doesn't work, well, bring on the statins. But for now, not with grapefruit juice. That restriction may change in the future, thanks to researchers at the University of North Carolina who have identified furanocoumarins as the troublesome compounds in grapefruit juice and have shown that they can be removed. The resulting juice had no effect on blood levels of medications. This research may yet have another benefit. Perhaps furanocoumarins can be added to medications, reducing the dose needed and reducing the risk of side effects. These developments should make grapefruit growers somewhat less sour.

BLUEBERRIES, ANTHOCYANINS, AND PTEROSTILBENE

L ife is sort of like walking a tightrope. We try to keep our balance as we struggle with diseases and aging, but we know that no matter what we do, eventually we will fall off. Obviously, anything that helps us stay on that rope longer is most welcome. And anthocyanins in blueberries may do just that. They sure do it for rats. At least that's what researchers at Tufts University in Boston found.

Why were these scientists interested in the unlikely combination of rats and blueberries in the first place? Because anthocyanins, the compounds responsible for the blue color of the berries, are powerful antioxidants. In fact, when different fruits and vegetables are tested for antioxidant activity, blueberries consistently rank near the top of the chart. And we know that antioxidants have been linked with inhibiting blood-clot formation, improving night vision, slowing macular degeneration, and generally reducing the risk of heart disease and cancer, as well as with protecting brain cells from aging. It is this anti-aging effect that captured the imagination of the Tufts researchers. First, a group of elderly rats was put on a blueberry-rich diet, while another group was treated to regular laboratory rat chow. Both groups were then exposed to high levels of oxygen for forty-eight hours. Inhaling extra oxygen produces copious amounts of free radicals. The idea was to study the effects of the free radicals on rats that had anthocyanins scurrying around in their bodies as compared with those that did not.

It is well known that free radicals can attack all sorts of molecules in the body, including those that play a role in the functioning of the nervous system. So it wasn't a great surprise that rats fed standard lab meals showed a significant impairment in neurological function when compared with the blueberry-treated rats. Apparently the anthocyanins were neutralizing the free radicals generated by the high dose of oxygen. But there was an even more important practical finding.

Rats enjoy walking on narrow ledges and beams, and are apparently very good at this practice—at least until they begin to age. Researchers can actually estimate the age of a rat by measuring the time it takes for the animal to lose its balance on a narrow beam. As rats reach the ripe old age of nineteen months, equivalent to between sixty-five and seventy years in a human, the average balance time drops from thirteen seconds to five. Older rats are also less adept at negotiating mazes, which of course is a real problem for a lab rat. But now comes the kicker. After eating blueberry extract for eight weeks, the old rats managed to stay on the rod for eleven seconds. They also negotiated mazes better! No great shock that the lay press seized upon this study, elevating blueberries to the status of a wonder food.

Of course there are no wonder foods. There are good diets and bad diets. And anthocyanins are present in various other fruits and vegetables as well. In fact, cherries are particularly rich in these compounds. Indeed, certain anthocyanins isolated from cherries have been shown to have anti-inflammatory properties, possibly useful in fighting arthritis. Even certain diabetics may respond favorably to anthocyanins in the diet. Dr. Muralee Nair at Michigan State University showed that in animal pancreatic cells, anthocyanins increased insulin production by some 50 percent.

The possible benefits of blueberries don't stop at anthocyanins. Recently researchers found that pterostilbene, another antioxidant present in blueberries, may reduce cholesterol. Interesting, but the study in question was not done on humans, and not even on live animals. It was done in the laboratory on rat liver cells. The researchers did manage to show that pterostilbene activates a specific receptor on these cells that is linked with reducing cholesterol and triglycerides. But nobody knows if this compound when ingested from blueberries acts the same way in a human liver, or indeed if it even gets there. Nobody knows how many blueberries would have to be eaten to lower blood cholesterol, but for sure eating a blueberry muffin or a blueberry-studded bagel is not going to do it. Nor will blueberry pancakes. Eventually we may find that anthocyanin or pterostilbene supplements in a pill form are useful, but for now I try to eat half a cup of blueberries several times a week. I'm not sure exactly how beneficial this is in terms of health, but I am sure of one thing: the berries taste better than pills.

CITRUS FRUITS AND SUPER FLAVONOIDS

Take some hamsters and feed them lots of fatty stuff to drive up their blood cholesterol. Then add some *super flavonoids* to their feed and hope that they mitigate the effects of the high-fat diet. Why? Because if the flavonoids reduce cholesterol, you've got a marketable product. Even better, since the super flavonoids derive from orange peel, you've got a "natural" product with all the associated commercial appeal. Well, in hamsters at least, the orange peel extract does deliver the goods. And in the future, it is certainly possible that some sort of standardized version of *polymethoxylated flavones* (PMFs) extracted from orange peels may be recommended for people with high blood cholesterol. But let's not jump the gun.

To most people, cholesterol is a dirty word. If you go by information garnered from the media, you might get the impression that a diagnosis of high cholesterol should prompt a swift meeting with an undertaker. While it is true that high cholesterol is a risk factor for heart disease, it is only one of many. High blood pressure, family history of heart disease, diabetes, lack of physical activity, and exposure to polluted air all play a role. In fact, half of all people who have heart attacks have normal or below-normal cholesterol levels. Nevertheless, a finding of high cholesterol certainly should trigger efforts to reduce it. But how? Physicians often reach for the prescription pad and scribble instructions for a statin drug. These are highly effective medications, but they come with side effects. Muscle pains and liver problems are possibilities—and the cost of the drug can be considerable. So people search for kinder, gentler therapies, hoping to find efficacy in "natural" products, which are perceived to be safer than synthetic drugs. The truth, of course, is that the safety of a substance does not depend on its source but on its specific molecular structure. Whether it was made by a chemist in a lab or by nature in a bush does not matter. What matters are the results of studies that have been carried out on the safety and efficacy of the substance.

There is no shortage of cholesterol-reducing claims on behalf of dietary supplements formulated from natural products. Some, such as guggulipids extracted from the guggul tree, seem exotic, while policosanol from sugar cane or allicin from garlic sound more mundane. Promoters of all such products muster up studies to hype their wares, but the scientific consensus is that the evidence is pretty thin. Still, the search for truly effective natural products continues, and justifiably so. After all, it is pretty clear that populations consuming more fruits and vegetables tend to have lower cholesterol levels. The question is whether this is due to what they are *not* eating, such as fatty meats, or to the presence of substances in plant products that actually lower cholesterol. The quest for such substances has resulted in the isolation of a variety of natural compounds that have been examined for possible cholesterol-lowering effects. Among these are the polymethoxylated flavones (PMFs), such as tangeretin, hesperidine, and naringin, which are found in citrus peel.

Why are these compounds of particular interest? For several reasons. First of all, PMFs are antioxidants and therefore have the potential of preventing the conversion of cholesterol into its more dangerous oxidized form. Also, studies in cultured cells showed that, like the statin drugs, PMFs inhibit the synthesis of cholesterol and triglycerides (fats in the blood) in the liver. Add to this the fact that Florida alone produces about 700,000 metric tons of orange-peel waste every year, providing ample raw material for PMF extraction, and you have potential for a profitable product. All you have to do is to show that it works.

Well, it certainly does in hamsters. When researchers at KGK Synergize, a Canadian company that specializes in studying natural compounds with disease-preventing or medicinal properties, carried out the hamster-feeding study, they found that incorporating tangeretin into the animals' diet reduced the levels of LDL (the famous "bad cholesterol") by as much as 40 percent. That's pretty exciting, at least for the researchers. And I suppose, for the hamsters. What, though, does this mean for us? We don't eat orange peel, but we could be encouraged to drink more orange juice, which also contains PMFs. But the problem is that we would have to drink twenty glasses of juice a day to get the amount of super flavonoids that the hamsters received. Incidentally, the term *super flavonoid* was coined by the KGK Synergize company researchers who appear to be very encouraged by their results. In fact, so encouraged were they that they have developed a commercial product, Sytrinol, which is a proprietary blend of citrus-peel extract and vitamin E.

The hope is that Sytrinol will prove to be an effective way to treat high blood cholesterol and, indeed, some preliminary human trials have produced optimistic results. In a study carried out at the University of Western Ontario, participants with elevated cholesterol were asked to take 300-milligram capsules (270 milligrams polymethoxyflavones and 30 milligrams vitamin E) of Sytrinol every day. After four weeks, there were highly significant drops in total cholesterol (20 percent) and in LDL cholesterol (22 percent), accompanied by an increase in HDL, the "good cholesterol." Unfortunately, only ten subjects were involved in the study, not enough to draw major conclusions, but certainly enough to stimulate bigger and better studies and more research. Especially given that citrus-peel extracts have also been shown to reduce insulin resistance in hamsters, implying that they may be of help in preventing diabetes.

And there is even some hope for cancer prevention. In the lab, tangeretin inhibits the growth of human mammary cancer cells, just like the widely used drug tamoxifen. In live animals, though, not only did the compound not offer protection, it neutralized tamoxifen's inhibitory effect. It seems then that women on tamoxifen therapy should not attempt to lower their cholesterol with citrus-peel products. For others, the flavonoids found in citrus products hold potential, although the adjective "super" to describe them appears to be somewhat premature.

ACAI BERRIES AND
ANTIOXIDANT POTENTIAL

The city of Belem in northern Brazil has a population of about two million people. It would be very interesting to find out if its inhabitants have an unusually low incidence of illnesses such as cancer, arteriosclerosis, or Alzheimer's disease. Why? Because the city is dotted with some 3,000 "acai points" where people line up to purchase a slurry made from the pulp of the fruit of the acai (a-sigh-yee) palm tree. Over 200,000 liters of the thick, purple sludge are consumed every day, which is more than the amount of milk that is drunk in the city. And, at least if you listen to some of the North American advertisers that have begun to import the juice of the acai berry, it has fantastic anti-inflammatory, antibacterial, antimutagenic, and, above all, antioxidant properties! "Nature's perfect fruit," boasts one distributor. "The Amazon's Viagra," brags another. Little surprise then that North Americans, in constant search for the next miracle that will help them beat the clock, are shelling out more than forty dollars for a bottle of juice made from the acai berry, which has "more antioxidants than any other edible berry on the planet."

One thing we can say for sure about antioxidants is that they help sell products. Just festoon a label with "source of antioxidants" and the food, beverage, or dietary supplement flies off the shelves. That's because researchers have shown that antioxidants can neutralize those nasty free radicals that form in our bodies as a result of breathing oxygen. And those free radicals have been linked with a number of human illnesses. So it stands to reason that if we can curtail their activity, we'll be better off. Fruits and vegetables are the main sources of antioxidants in our diet, and the prevailing opinion is that it is their antioxidant content that is responsible for the health benefits seen in populations with a high intake of plant products. But trials using antioxidant supplements have repeatedly failed to show the expected positive results. Fruits and vegetables contain dozens of compounds that have potential physiological activity, and it seems that it is a blend of these that is required for health benefits. In other words, the whole is somehow greater than the sum of its parts.

There is no doubt that antioxidants from fruits and vegetables are an important part of our diet, but the relevance of a single food or drink having more or less of these compounds is questionable. What matters is our total antioxidant intake. On a weight basis, acai berries may have a higher concentration of antioxidants than apples, but it is certainly easier to load up on apples. However, there is another important issue. The *antioxidant potential* of a food is determined by a laboratory measurement. One common method relies on generating free radicals by means of a chemical reaction in the presence of alpha-keto-gamma-methiolbutyric acid (KMBA)— how's that for a tongue twister! Free radicals attack KMBA, break-ing it down and releasing ethylene gas in the process. Ethylene can then be identified and quantified by an instrumental technique known as gas chromatography. Adding a food extract containing antioxidants to the mix neutralizes free radicals and therefore re-duces the amount of ethylene gas released, providing a measure of antioxidant potential.

It is such measurements that fuel the claim of acai berries being a particularly good source of antioxidants. However, a laboratory flask is a far simpler system than the human body. We don't know how well the antioxidants in a given food are absorbed into the bloodstream, and we don't know if they have the same neutraliz-ing effect on free radicals in the complex environment of the body as they have in the lab. And we certainly don't know that what-ever activity they have is enough to prevent any specific disease.

The only way to know that is by means of a controlled trial: give a large group of people a regular dose of acai juice while another similar group takes a placebo, then follow them for years and monitor disease patterns. Nobody has done this; therefore any health claim for acai is pure conjecture.

Of course, this doesn't mean that the possible health-promoting properties of acai berries should not be investigated further. Any food with a high antioxidant potential merits investigation. A recent study at the University of Florida, for example, showed that acai-berry extracts destroyed a high percentage of leukemia cells in culture dishes. Interesting, but not all that unusual. Extracts of mangoes and grapes do the same. In any case, this is a long, long way from showing that such extracts have any effect on leukemia cells in the body. But such studies are enough to supply the ammunition that some unethical marketers use to hype the "anticancer" effect of acai juice. Maybe they need to learn a lesson from the promoters of Xango, a mangosteen juice product that was all the rage before the company received a warning letter from the US Food and Drug Administration (FDA). Claims of antitumor benefits, lowering of blood pressure, and prevention of hardening of the arteries are not supported by science, said the letter. In fact such claims can only be made on behalf of a drug, and as such, the product would require FDA approval, which is contingent on providing supporting evidence. Again, this is not to say that compounds in mangosteen, such as the much talked-about xanthones, may not eventually turn out to have health benefits. But claiming that the juice can ward off disease is unproven, at best.

The chance that mangosteen or acai juice can make a significant contribution to our antioxidant status is slim. Better to concentrate on getting five to ten servings of common fruits and vegetables every day. Where acai berries have a real potential, though, is in helping the economy of Belem, where approximately 110,000 metric tons of fruit are worked up commercially every year, leaving behind 100,000 metric tons of seeds. These, like the fruit, have high antioxidant potential but little commercial application. Perhaps extracts can be used as preservatives in foods, and it may even turn out that concentrates have a therapeutic potential. But if that turns out to be the case, you will hear about it from the *New England Journal of Medicine*, or some other such peer-reviewed publication, and not from your neighbor who has become involved in selling acai juice through a multilevel marketing scheme.

FISH AND OMEGA-3 FATS

"The sardines, Jeeves, eat the sardines!" It is with these words that Bertie Wooster, in P. G. Wodehouse's beloved stories, implores his quick-witted gentleman's gentleman to rev up his mental engine to extricate his master from yet another romantic jam. Jeeves always rises to the occasion and comes up with some clever scheme to deliver Bertie from his predicament. Whether Jeeves actually makes use of the advice to load up on fish is unclear, but Wodehouse's repeated references to fish consumption and brain power attest to the widespread nature of this common belief. Can eating fish really make us smarter? Maybe.

The first attempt to put the long-held notion that "fish is brain food" on a scientific footing emerged during the eighteen hundreds when scientists discovered that the key molecule in producing cellular energy, namely adenosine triphosphate, or ATP, was rich in phosphorus. Since ATP provides the energy for thinking and is used up in the process, these scientists proclaimed that its regeneration was the key to mental acuity. Since fish was an excellent source of phosphorus, it stood to reason that it was "brain food." Today, researchers know that there is no dietary shortage of phosporus, so this is not the case. But, interestingly, other compounds in fish may play a role in brain function. Two particular fats, docosahexaenoic acid (DHA) and eicosapentaenoic acid (EPA) merit investigation. These are the celebrated omega-3 fats that have also been linked to a reduced risk of heart disease. While fish oils, like vegetable-derived oils, have several carbon-carbon double bonds in their molecular structure, one of these always involves the third carbon from the end of the molecule. That end carbon is termed the *omega* carbon, from the Greek word for *end*; hence the name *omega-3 fats*.

The human brain is composed of about 60 percent fat. But it seems that it is the composition of brain tissue in terms of specific types of fats that is the key to predicting mental prowess. The earliest research suggesting such a connection focused on monkeys. When these animals are fed a diet deficient in DHA, their brains and eyes do not develop properly. This is not all that surprising, given that DHA is the primary fat found in the brain and in the retina of the eyes. Interestingly enough, supplementing the diet with DHA restores normal brain and eye development in the monkeys, demonstrating that the composition of the brain responds to dietary intake.

What about humans? We're often told that we are what we eat. Do we also think with what we eat? Some interesting evidence emerges when epidemiologists examine rates of depression around the world. The variation is surprising, the incidence being sixty times greater in some countries than in others. Unfortunately, Canada and the United States are at the high end, while countries such as Korea and Japan have a very low incidence of depression. When fish consumption is brought into this picture, a remarkable relationship appears. Countries with high fish intake have low rates of depression and countries with low intakes exhibit high rates. Furthermore, studies have demonstrated a link between the increase in depression in North America and the decline in the consumption of foods rich in DHA. Obviously these observations do not necessarily mean that eating fish can reduce the risk of depression, but there does appear to be some corroborating evidence to suggest this conclusion.

Low concentrations of a chemical found in the cerebrospinal fluid, 5-hydroxy-indolacetic acid (5-HIAA), have been quite conclusively linked with depression and suicide. We also know that people with low levels of DHA in their blood plasma have low levels of 5-HIAA.

Interesting. Then consider that researchers at the University of Surrey, as well as at Purdue, have linked low blood levels of DHA to dyslexia, attention deficit disorder, and hyperactivity, and have shown an improvement in these conditions using a DHA supplement now marketed as Efalex. Furthermore, a study of more then one thousand elderly people who were followed for nine years showed that those with high blood levels of DHA were more than 40 percent less likely to develop dementia, including the Alzheimer's variety. This disease is caused by the buildup in the brain of a protein known as amyloid. When mice genetically engineered to develop Alzheimer's disease are fed food fortified with DHA, they form significantly fewer amyloid deposits in the brain.

Add to this the results of a Japanese study that demonstrated improved short-term memory and night vision in healthy subjects taking DHA supplements, and those of a Dutch study that showed that cognitive impairment and decline in elderly men was inversely associated with fish consumption, and a fairly consistent picture emerges. Healthy brain function requires adequate levels of dietary DHA. If we are looking for even more evidence as to the importance of this particular fat in the diet, we need look no further than our very first meal: breast milk is a particularly concentrated source of DHA, probably an evolutionary reflection of the importance of this fat in the eye and brain development of infants. Indeed, as more and more information about the importance of DHA accumulates, infant-formula manufacturers are focusing on adding it to their product.

DHA and EPA may not only oil our brains, they may also protect our hearts. These days doctors often recommend that cardiac patients supplement their diet with fish oil, the World Health Organization (WHO) urges people to eat fish a couple of times a week, and products fortified with omega-3 fats are mushrooming on supermarket shelves. Many scientific studies justify such optimism about omega-3 fats, but curiously, there are also studies that find no beneficial effect. That's why a group of British scientists led by Lee Hooper of the University of East Anglia went fishing in the deep sea of scientific studies and netted a controversy. They decided to scour the literature for the best studies and pool the results with the hope of coming up with solid evidence-based recommendations about omega-3 fat intake. After looking at over 15,000 papers, they focused on eighty-nine studies that they determined to be the most meaningful. More than half of these were randomized, controlled trials, meaning that participants were randomly given either a placebo or a fixed dose of omega-3s, while the others were cohort studies in which populations with varying omega-3 intakes were followed and their health status analyzed. The results were surprising. Why? Well, to understand that, we need a little background.

By the 1970s, scientists had gathered enough evidence about a link between fat in the diet and increased risk of heart disease to urge people to cut back on fatty foods. But there was a conundrum. The Inuit who lived in the far north and subsisted almost entirely on fatty fish had a remarkably low incidence of coronary artery disease. Similar effects were noted in other high fish-consuming cultures, such as the Japanese. Could there be something special about the type of fat found in fish? Certainly there was a difference in molecular structure. Perhaps the omega-3 fats had different metabolic pathways from other fats and a different impact on health. There was only one way to find out.

Epidemiological studies gave the first clues. Many, but certainly not all, showed that people who eat more fish have a lower risk of cardiovascular disease. In a typical case, researchers who followed employees of the Western Electric Company in Chicago found that men who consumed on average thirty-five grams (1.2 ounces) or more fish daily had a significantly reduced risk of heart disease. Then came the intervention trials. In one such study in India, patients admitted to a hospital after a heart attack were given a daily dose of 1,800 milligrams of fish oil or a placebo. After one year, 35 percent of the patients in the placebo group had a "cardiac event," while the fish-oil group had only a 25 percent incidence.

Interesting, but not highly significant. A similar Norwegian study showed no such benefit even though the dose of fish oil was higher. Norwegians, though, have a high fish intake, so perhaps they already had maximized the protection available from fish oils. Other studies showed that patients who had blocked coronary arteries as seen in angiograms did better if they were supplemented with 3,000 milligrams of fish oil a day, but those who had their arteries opened up through balloon angioplasty did not benefit even from high doses of fish oils.

What's the theory behind the possible benefits of fish oil? Omega-3 fats can act as anticoagulants and reduce the chance of blood-clot formation, which means a reduced risk of heart attack. They also have anti-inflammatory properties, and inflammation is being increasingly implicated in many diseases, including heart disease. But the biggest benefit likely comes from fish oil's ability to prevent irregular heartbeat. In animals, the effect of drugs that induce irregular heartbeat can be countered by the presence of omega-3 fats in the blood. This finding has been corroborated in humans as well. A British trial, for example, showed that heart patients advised to eat two servings of oily fish a week, or to take daily fish-oil capsules for two years, had a significantly lower death rate than patients who were told to increase their fiber intake and reduce fat consumption. An Italian study of over 2,800 heart attack survivors also showed that fish-oil capsules providing 850 milligrams each of EPA and DHA dramatically reduced the incidence of death in the first nine months following a heart attack. The protection, however, seems to fade with time, even if fish-oil consumption is maintained.

But to confuse the situation, patients with implanted defibrillators have not shown a benefit when given fish-oil supplements. In fact, in one study, they showed an increased risk. Then there is the Physicians' Health Study, which has followed some 17,000 male physicians over many years, and found that men who ate fish more than five times a week had an increased rate of atrial fibrillation, a type of irregular heart beat. Certainly, more fish is not better!

Obviously, there are inconsistencies in the results of studies that have explored the relationship between heart disease and fish consumption. And it is precisely for this reason that Lee Hooper and colleagues attempted to look at all the relevant studies and distill out information that could form the basis of recommendations to the public. Now for the surprise. After their thorough analysis, the only conclusion the researchers could come to was that omega-3 fats do not have a clear effect on cardiovascular events or death rates.

It is hard to know what to make of this. Why do some studies show potent protective effects attributed to fish consumption while others, seemingly equally well done, turn out to be disappointing? Maybe fish oils have different effects in people depending on their age or state of health. Maybe the benefits of fish are not in what they contain, but in what they displace from the diet. Better to eat fish than hamburgers and trans fat–laden fries.

So what is our bottom line here? Two fish meals a week with emphasis on salmon, sardines, herring, and mackerel are likely to do more good than harm. Why a question of harm? Because fish can be contaminated with methyl mercury, PCBs, and dioxins, all of which can present a risk to health. That's why pregnant women and young children are advised to avoid shark, swordfish, fresh and frozen tuna, king mackerel, and tilefish, the ones most likely to be contaminated, and limit others to about 350 grams (twelve ounces) a week. Eating this much is a great idea. A study in *The Lancet* showed that pregnant women who consumed about 300 grams of fish or seafood a week gave birth to children with higher tested IQ scores.

Middle-aged men and postmenopausal women, for whom the benefits of eating fish greatly outweigh the risks, do not have to watch their intakes so carefully but should still limit their intake of fresh or frozen tuna, shark, or swordfish to 1,000 grams per week. Canned tuna is low in mercury, particularly if it is "light." Salmon, especially the wild variety, is also low in mercury and other contaminants. Canned Alaska salmon is a good choice.

People with existing heart disease fall into a different category and should speak to their physicians about increasing their intake of omega-3 fats to about a gram a day. This amount can be had from between 60 and 90 grams of sardines, 60 grams of herring, 100 grams of salmon or trout, or 500 grams of cod. The omega-3 fat content of tuna is variable, and up to 300 grams may be needed to provide a gram of omega-3 fats. For days when fish cannot be consumed, supplements may be appropriate. Most common capsules provide 180 milligrams of EPA and 120 milligrams DHA, so several may be required. These, though, can present problems. First, there is the issue of fragrance. One breath and a thousand cats drool. There is also the possibility of nausea and gastrointestinal disturbances. Still, the benefits are likely to be worthwhile. And the benefits may not be restricted to the brain or the heart. Several studies have shown that eating fish twice a week can reduce the risk of macular degeneration, the leading cause of age-related blindness.

FLAX, OMEGA-3 FATS, AND LIGNANS

Fish are not the only source of omega-3 fats. Oils derived from flax, canola, and soybeans are rich in alpha-linolenic acid (ALA), which, although not identical to DHA and EPA, has similar biological activity. So it too can have an effect on mental acuity and heart disease. Furthermore, some ALA is converted to DHA and EPA in the body. Therefore, results from the Nurses' Health Study, monitored by Harvard University researchers, are not surprising. This study has followed the health status of over 76,000 nurses who, starting in 1984, filled out food questionnaires every four years. We now learn that women who consumed the most alpha-linolenic acid from foods had a 46 percent lower risk of sudden cardiac death than women who consumed the least. The major sources of ALA were leafy green vegetables, walnuts, canola oil, and flax.

Flax oil is a great source of alpha-linolenic acid, but flaxseeds offer more than just ALA. A couple of thousand years ago, Hippocrates, the famous Greek physician, was already aware of the benefits of flax. His prescription for patients who suffered abdominal pains was simple: "Let them eat flax!" And it's probably not bad advice—as long as the pain stems from constipation. It turns out that flaxseeds, which come from the plant used to make linen, are an excellent source of dietary fiber. This indigestible plant component provides a laxative effect by allowing wastes to absorb water as they pass through the digestive tract. And the soluble portion of fiber provides an added plus. On its journey through the digestive tract, it binds cholesterol and prevents it from being absorbed. Furthermore, it also binds the bile acids needed for digestion, forcing the body to produce more. Since the starting material for the biosynthesis of bile acids is cholesterol, blood levels of cholesterol go down.

We don't even have to eat flaxseeds directly to get some of the benefits of alpha-linolenic acid. How about some omega-3 eggs? That sounds paradoxical, because when eggs are mentioned, the first word that often comes to mind is *cholesterol*, which in turn conjures up thoughts of clogged arteries and premature demise. In truth, blood cholesterol responds much more to the saturated fats found in meat and full-fat dairy products than it does to cholesterol in egg yolk. Still, eggs suffer from an image problem. Omega-3 fats, on the other hand, positively bask in the limelight these days. Slipping these fats into eggs would certainly be a healthy boost to their image. Especially considering that many people worry about pollutants like mercury and PCBs, both of which crop up in fish.

How do you then enrich eggs with omega-3 fats? You could feed the chickens fish meal, but that would make the eggs taste fishy. Or you could feed them flaxseed, which of course is a good vegetable source of omega-3s. This results in each egg having about thirty-five milligrams of ALA and thirteen milligrams of EPA and DHA, not an insignificant amount. Eating five to seven of these eggs a week is roughly equivalent to one serving of fish. Of course, for someone who consumes very few eggs, switching to the omega-3 version offers no benefit.

There is more to flaxseeds than offering protection against heart disease. They are a source of lignans, compounds that have anticancer properties. Lignans are actually not found in flax, but they form in our colons where bacteria feed on secoisolariciresinol, a precursor that is found in flax. We were first alerted to the potential benefits of lignans by Finnish researchers who observed lower levels of these substances in the urine of women with breast cancer than in the urine of healthy women. This was an interesting finding, especially given that lignans have a chemical similarity to estrogens. (Since they originate from a plant source, they are termed *phytoestrogens*.) The majority of breast cancers are estrogen positive, meaning that the irregular multiplication of cells that characterizes the disease is triggered by the body's own estrogen. Could it then be that lignans, with their resemblance to estrogen, fit into estrogen receptors on cells and block estrogen from stimulating cellular activity? Sort of like the wrong key that fits a lock, unable to turn but able to prevent the right key from entering.

A nice theory, but it requires experimental evidence to back it up. And Dr. Lilian Thompson at the University of Toronto attempted to provide it. She fed varying amounts of flaxseed to animals that had been treated with a chemical known to produce estrogen-sensitive tumors. To her satisfaction, she found that the animals fed flax produced fewer and less-aggressive tumors. But tumor reduction was not the only effect observed. In female rats, small doses of flax delayed puberty, larger doses brought it on sooner. In male rats, 5 percent flax in the diet reduced the proliferation of cells in the prostate, but 10 percent increased it. Such results are somewhat worrying, especially considering a study back in 1994 that linked high blood levels of alpha-linolenic acid with an increased risk of prostate cancer. This was a surprise because previous studies had shown a decrease in prostate cancer risk with increased fish-oil consumption. Was ALA somehow different from other omega-3 acids in its effects on the prostate? The situation is confusing because tissue taken from the prostate gland of cancer patients does not show higher levels of ALA, even though such levels are found in the blood plasma. In any case, dietary flax has not been linked to any human health problem. Quite the contrary.

Consumption of flaxseeds has been associated with a reduced risk of both breast and prostate cancer. Indeed, Lilian Thompson has shown that women awaiting surgery for breast cancer benefit from daily consumption of muffins fortified with twenty-five grams (less than one ounce) of ground flaxseed. And Duke University researchers found that three tablespoons (forty-five milliliters) of ground flax daily given to men waiting for prostate cancer surgery improved the chances for a successful outcome. Levels of testosterone, a hormone that can stimulate cancer cells, were lowered and there was an observable decrease in cancer cell proliferation.

Let's summarize what we know about flax and ALA, the specific omega-3 fat found in the seeds. ALA protects against heart disease, perhaps by reducing inflammation or by curtailing irregular heartbeat. Flax is a great source of soluble fiber that binds bile acids in the gut, forcing the liver to produce more. The raw material it needs for bile-acid production is cholesterol, so the end result is that flax can lower blood cholesterol. The fiber in flax also slows the rate at which glucose is absorbed from food, and diabetics have indeed seen a drop in blood glucose with regular flax consumption, in some cases by as much as 30 percent in response to a daily intake of fifty grams of flaxseed. And of course we also have the cancer-preventive effect of the lignans. On the dark side, there is the lingering potential connection between flax oil and prostate cancer. But the benefits outweigh the risks. A heaping tablespoon of ground flax sprinkled on your oat bran in the morning, which of course is then topped off by some berries, is a good idea.

CANOLA AND ALPHA-LINOLENIC ACID

What is canola? To many it's a mystery. They don't know if you hunt it, fish it, or grow it. But they know that somehow "canola" can be used to produce cooking oil. And, as is often the case with foods of a somewhat baffling origin, questions arise in people's minds, particularly about health effects. Let's cut to the chase. Canola is a plant that produces seeds that can be pressed to yield oil. It is one of the best, safest, and most economical oils that can be used in food preparation.

Now, that is not exactly what you may have heard or read before, is it? You may have heard that the name *canola* was invented to distract consumers from the fact that the oil actually derives from the toxic rapeseed plant, which has been linked to glaucoma, respiratory problems, neurological diseases, and malfunctions of the immune system. And to top it off, you may even have heard that canola oil is the source of the notorious chemical warfare agent, mustard gas. The source of all this claptrap is an e-mail that has been widely circulating since about 2001 and seems to gather more nonsense with each go-around. The latest gem recounts the saga of a woman whose arm was "slightly banged and split open like it was rotten." She called her mother to ask what could have caused her injury. (One would think that if your arm splits open your first reaction would be to go to the hospital.) In any case, the astute mother remarked, "I'll bet anything you are using canola oil!" And sure enough, we are told, there was a big gallon jug in the pantry. Is it possible that some people actually believe such hogwash? Judging by the questions I've gotten on this issue, the unfortunate answer is yes.

Attempts to trace the origin of the stunning misinformation in the anticanola e-mail always end at the same place: John Thomas in his 1994 book, *Young Again*, in which the author claims to have reversed his "bio-electric age" (whatever that may be) by eliminating the likes of canola and soy oil, using liver cleansers (which he sells), drinking specially filtered water (which he sells), and taking dietary supplements that are specially tuned to the "frequencies" of his body. Readers can get in on this too by sending Thomas a picture of themselves, which will be analyzed with a special machine (which he has) to determine the sender's "personal frequencies," the results of which can then be used to customize appropriately tuned dietary supplements (which he sells). And what qualifications does this remarkable man who "was encouraged to write *Young Again* because he does NOT age" actually have? As far as I have been able to find, none! Aside from a nondescript picture on the book's back cover, I can't find hide nor hair of John Thomas. It is amazing, though, how a vacuous nobody has been able to make so many people twitch about the safety of canola oil.

About the only thing Thomas got right in his silly diatribe is that *canola* is indeed a name that was coined for a special variety of rapeseed. It comes from a clever combination of the words *Canada*, *oil* and *low acid*. Rapeseed oil has long been used as a lubricating oil, but its somewhat bitter taste, due to compounds called glucosinolates, impaired its use in food. There was also an issue about another component, erucic acid, which in some animal studies, when incorporated into the diet in grotesque amounts, caused fatty deposits in several organs. Last century, Canadian researchers, using traditional plant-breeding methods, managed to develop rapeseed with a low glucosinolate level and minimal erucic acid content. The oil pressed from the seeds of these plants became canola oil.

Like all oils, canola is composed of three fatty acids linked to a backbone of a glycerol molecule. Both the cooking performance and health properties of fats and oils are determined by the types of fatty acids they contain. Saturated fats, containing no carbon-carbon double bonds in their structure, are implicated in heart disease but can be repeatedly heated when it comes to frying. Monounsaturated fats, with one double bond, and polyunsaturated fats, with many, are more heart healthy, but are less stable to heat. As we have seen, some polyunsaturated fats, such as alpha-linolenic acid (an omega-3 fatty acid), have been specifically linked with protection against heart disease. It turns out that of all the commonly used oils, canola has the lowest content of saturated fats and, next to flaxseed oil, the second highest content of alpha-linolenic acid. Actually, one of the best ways to judge the health properties of a fat, aside from being low in the saturated variety, is the ratio of omega-6 to omega-3 fatty acids. (The numbers refer to particular positions of the double bonds in the molecular structure.) Canola oil has the ideal ratio of 2:1.

Because canola is so high in unsaturated fats, it does not stand up well to prolonged heating as is required in restaurant frying and does not have the keeping qualities desired by food production industries. Hydrogenating the oil makes it more suitable, but also introduces trans fats. Indeed, it is a good idea to minimize intake of hydrogenated fats, whether these come from soy, corn, canola, or any other oil. As far as home use goes, though, nonhydrogenated canola oil is a great all-purpose oil. Incidentally, there is no truth to the rumor that heating unsaturated oils produces trans fats. Heating foods to a high temperature does, however, produce some nasty compounds widely regarded as carcinogens. That's why any sort of frying should be limited. I use canola oil to fry my Wiener schnitzel, but I use the oil only once, and of course do not indulge in this delicacy very often. But when I do, I have no concern about being deprived of my life force *chi*, being poisoned by cyanide, or having my brain damaged à la mad cow disease, all of which, at least according to the witless John Thomas, are consequences of canola consumption. Actually, judging by his example, it seems that it is avoiding canola oil that causes brain damage.

OLIVE OIL AND OLEIC ACID

Sometime around 1500 BC, the island of Crete was shaken by a huge earthquake. Trying to placate the gods of the underworld who were shaking the earth, the natives lowered a supply of their most precious food into a deep well. That is how, in 1960, archaeologists came to discover a large bowl of olives, well preserved by the cool temperatures, at the bottom of an ancient Cretan well. Did the inhabitants of Crete way back then know something about the health benefits of olives? Hard to say, but their descendants certainly seem to be a healthy bunch. At least that's what famed epidemiologist Ancel Keys found in the 1960s when he examined disease patterns in different countries and attempted to relate these to lifestyle factors. Crete proved to be especially interesting because people there seemed to have particularly long lifespans and low rates of heart disease and cancer, and they routinely worked well into old age. Keys learned that the Cretan diet had been essentially the same for centuries, with its most basic element being virgin olive oil. That in itself did not mean much, as Keys realized, but he became more intrigued when he discovered that Cretans who had migrated to the United States suffered the same heart disease and cancer rates as the rest of the American population. Could diet be a factor?

When Keys analyzed his data further, a pattern began to emerge. Countries where a lot of saturated fat, as in meat and dairy products, was consumed, had high rates of heart disease, whereas countries where liquid vegetable oils served as the main source of fat had a decidedly lower incidence of the disease. An explanation for this observation materialized when researchers discovered that heart disease rates were linked to the amount of cholesterol in the bloodstream, and that cholesterol levels in turn were determined by the fat composition of the diet. The critical feature seemed to be whether or not the fat molecules contained carbon-carbon double bonds in their structure. Those that did, the unsaturated fats found in vegetable oils, lowered cholesterol while the saturated fats devoid of such bonds raised it.

As physicians learned of this relationship, they began to advise the public to alter their eating habits. Butter and lard were out, vegetable oils were in. And heart disease rates began to plummet. But you can't spread vegetable oil on bread and you can't make flaky pastry with it, so a compromise seemed to be in order. And the food industry knew just what to do. Producers geared up to react polyunsaturated fats with hydrogen gas to produce "partially saturated" solid fats, which were to be kinder to our coronary arteries than the infamous saturated ones. Margarine and vegetable shortening, widely promoted as containing less saturated fat than butter, became dietary staples. Only later would we learn that hydrogenation introduces the notorious trans-fatty acids, which, although still unsaturated, just may be worse for us than saturated fats. Hydrogenation thus blurred the link between consuming unsaturated fats and health benefits. Too bad, because those benefits are real, as long as we are talking about unsaturated fats without trans-fatty acids. Like olive oil.

Olive oil is mostly a monounsaturated fat, which chemically speaking means it has one carbon-carbon double bond in its molecular structure. Saturated fats, as we have seen, are linked with heart disease, and there is also some concern about a diet that is very high in polyunsaturates, since in animal studies they promote the growth of colon and breast tumors. There is a probable rationale for this finding: polyunsaturated fats are more likely to give rise to free radicals, which have been linked with all sorts of things ranging from cancer to aging. As far as monounsaturated fats are concerned, though, the news is good, especially when it comes to olive oil. While it doesn't lower cholesterol better than polyunsaturates like corn or soy oil, it may offer benefits that these don't, such as protection against cancer.

Back in 1995, the lay press got all excited about a study carried out in Greece implying that women who consumed olive oil more than once a day were able to reduce significantly their risk of breast cancer. Actually, the study was poorly carried out and relied on a single questionnaire to estimate dietary intake over the preceding year, a notoriously unreliable technique. But the study did have a significant effect. It stimulated more research into the possibility that olive oil offers protection against cancer. At Northwestern University, for example, researchers treated human breast cancer cells with oleic acid, the main monounsaturated fat in olives. The dose used was akin to that found in the bloodstreams of people whose diets are rich in olive oil. Interestingly, the acid halved the production of a protein known as HER2/neu, which plays a key role in about a fifth of all breast cancers. But oleic acid may not be the only anti-cancer substance in olive oil. A study at the University of Ulster has revealed that specific phenols found in virgin olive oil reduced the damage to DNA in cultured colorectal cells. Of course, this is just a laboratory finding, but it does mesh with the observation of lower rates of colorectal cancer in Mediterranean countries where olive oil is widely consumed. Add to this the recent finding that extra-virgin olive oil contains oleocanthal, an anti-inflammatory substance with pharmacological activity similar to that of ibuprofen (e.g., Advil), and olive oil begins to look even more attractive.

Many restaurants now recognize the appeal of olive oil and offer little dishes of it instead of butter on the table. The key phrase is "instead of." Adding a couple of spoonfuls of olive oil a day to the diet is not the way to go; substituting it for saturated fats is. And if you want a further testimonial, just ask Madame Jeanne Calment.

Actually you can't. That's because she died in 1997, at the age of 122, the oldest person ever to live. She attributed her longevity to port wine and olive oil. Mme Calment also rubbed the oil on her skin and once quipped, "I've never had but one wrinkle, and I'm sitting on it."

SOY PROTEIN AND SOY ISOFLAVONES

I never thought I would get excited about tofu. Frankly, the taste doesn't thrill me. But I am intrigued by some of the scientific studies that have linked eating soy products to protection from disease. Japanese women, for example, have one-quarter the breast cancer rate of North American women. And Japanese women eat a lot of soy products. This does not necessarily mean that soy consumption has anything to do with breast cancer. However, it may well be that the soy connection is more than just a chance association. Before exploring the possible benefits of soy, though, let's remember that breast cancer is a complex disease with many possible contributing factors. It is age related, has a genetic component, and is also linked to excessive alcohol consumption. There may be a connection to high levels of certain fat-soluble pesticides, and the types of fat in the diet may also play a role.

Our soy saga really starts back in the 1940s when Australian farmers noticed that sheep grazing on a certain type of clover failed to reproduce normally. Veterinarians found that the animals' urine contained high levels of equol, a compound previously found in the urine of pregnant horses. As it turned out, bacteria in the sheeps' intestines were capable of converting a naturally occurring compound in clover to equol. And equol was known to have biological activity similar to estrogen. It was no great surprise that an estrogen-like substance should interfere with fertility; after all, estrogen was known to play an important role in human reproduction. Scientists then began to wonder whether other plants might also produce compounds with estrogenic activity. Enter the soybean. This Asian staple turned out to be rich in phytoestrogens (plant-derived estrogens) known as isoflavones. Genistein and daidzein, in particular, were of interest because they were partially excreted in the urine and could be correlated with the amount of soy in the diet.

The discovery of phytoestrogens raised eyebrows because scientists already suspected that estrogen and breast cancer were somehow connected. Women exposed to more estrogen over a lifetime were known to have a higher risk of contracting the disease. This includes women who come into puberty early, reach menopause late, or have few or no children. In other words, it seems that any factor that lowers the total number of menstrual cycles over a lifetime lowers the risk.

Now we get back to our Japanese women. They have longer menstrual cycles, averaging thirty-two days compared with the North American twenty-nine days. This could mean thirty to forty fewer periods in a lifetime. They also have up to 1,000 times more phytoestrogens in the urine than North American women. But the soybean plot really thickens when we note that Japanese women consume thirty times more soy products than we do, and that Japanese people who migrate to North America and take up the North American diet and lifestyle show cancer rates comparable to those of other Americans.

We can even postulate a possible mechanism for the isoflavone–breast cancer connection. Some cells in breast tissue are known as estrogen responsive, meaning that they contain certain proteins (estrogen receptors) to which estrogen can bind, very much in the fashion of a key fitting into a lock. This binding unleashes a sequence of events in the nucleus of the cell that eventually leads to the manufacture of certain proteins that cause cell proliferation. It is such abnormal cell multiplication that leads to cancer. Isoflavones, it seems, are actually "weak" estrogens. They fit into estrogen receptors but do not stimulate much cellular activity. At the same time, they prevent estrogen from binding with the receptor. It is as if the wrong key had been inserted into the lock. The key cannot be turned, but it effectively prevents another key from being inserted.

So much for associations and theory. What practical evidence can we muster to show that soy may actually prevent breast cancer? A number of animal studies have demonstrated that the consumption of soy or isolated isoflavones reduces tumor development. Harvard researchers found that rats fed isoflavones for two weeks before being injected with breast cancer or prostate cancer cells developed far fewer tumors than control rats. Animals that drank tea in addition to the isoflavones did even better. Human data are less direct, but some do exist. Dr. David Jenkins of the University of Toronto examined urine from volunteers on a low-fat diet that included thirty-three grams (just over one ounce) of soy protein a day. The diet resulted in reduced hormonal activity in the urine as measured by the effect on human breast cancer cell lines. Jenkins suggests this corresponds to a slight protective effect against breast cancer.

Researchers have also compared groups of breast cancer patients with matched controls and noted a decreased risk of up to 50 percent in premenopausal women who consumed soy daily. A classic study in Singapore showed that breast cancer rates correlated inversely with the amount of soy protein eaten on a regular basis. More than twenty studies of Asian women have shown that even one cup (250 milliliters) of soy milk or half a cup of tofu a day is associated with reduced cancer risk. In addition, some studies have found that menopausal women who start eating twenty grams of soy protein powder daily (roughly equivalent to a soy burger, a cup of soy milk, or a serving of tofu), show a reduction in the severity of menopausal symptoms. An added benefit is increased bone density in the spine. As far as premenopausal women go, the same kind of diet increases the length of their menstrual cycles by 2.5 days, while the isoflavone content of their urine also rises dramatically. It is certainly apparent that soy has estrogen-like activity!

Genistein, the main isoflavone, may have yet another effect. It decreases the growth rate of blood vessels that nourish tumors. This inhibition of what is called *angiogenesis* may turn out to be an important anticancer effect. It may even explain why men who have high levels of genistein in the urine seem to be protected from prostate cancer. Although the isoflavones appear to be the most interesting anticancer compounds in soybeans, there are others. Folic acid, for one, has been shown to prevent mutations in DNA.

There are, however, some inconsistencies in the soy saga. A Japanese study showed that women with breast cancer had consumed no less soy than a control group unaffected by the disease. Chinese women, who eat only about a third of the soy-based foods that the Japanese eat, have the same low rate of breast cancer. Of course, it is possible that a certain amount of soy is protective but eating more carries no further benefit, and may even present a risk. To add to the confusion, in the test tube, at very low concentrations, genistein enhances human breast cancer cell proliferation while at higher concentrations it inhibits it. The timing of soy consumption may also be important. Female rats, for example, are protected against carcinogen-induced breast cancer if soy is given before puberty, but not if given later in life. In humans, isoflavones may act differently after menopause when natural estrogen levels are lower, than earlier in life when copious amounts of estrogen are produced. Epidemiological studies of Asians do indeed show that it is soy intake early in life that is protective. Later on, at least in theory, when there is less competing natural estrogen, soy may have a different effect.

After menopause, women produce varying amounts of estrogen and even small variations may be important. If estrogen levels are low, high doses of soy can have an adverse estrogen-like effect, but soy's isoflavones can block the negative effects of natural estrogens if these are present at a high level. This is more than just a theoretical possibility. Charles Wood at Wake Forest University treated postmenopausal monkeys with a high or low dose of estrogen and then fed them diets with varying amounts of isoflavones. Isoflavones had no effect on the low-estrogen animals, suggesting that soy intake after menopause does not increase breast cancer risk. Even more comforting were the results seen in the animals given a high dose of estrogen. When these were put on a diet containing 240 milligrams of isoflavones a day, they showed a reduced susceptibility to breast cancer! It is hard to know what to do with this information because that amount of isoflavones is attainable only from supplements, but at least it reduces the concern that phytoestrogens might mimic estrogen when it is not around. It seems that postmenopausal women who want to give soy products or isoflavone supplements a try to reduce menopausal symptoms need not worry about increasing their breast cancer risk. While many studies have come up short, some have shown that a daily intake of about 160 milligrams of isoflavones can lead to a reduction in hot flashes and night sweats.

While the available information suggests that soy, particularly if consumed early in life, may protect against breast cancer, it doesn't shed light on soy's effects when the disease is already present. There just isn't enough data to make recommendations in this case, but the prudent approach would be to refrain from going overboard with soy consumption.

Of course, everyone worries about cancer, but heart disease is a bigger killer. And here too there is a lot of talk about soy's protective effects. Actually, there is more than just talk. By 1999, enough scientific evidence had been gathered by soy product producers to petition the US Food and Drug Administration to allow them to put a "heart healthy" claim on their packages. After reviewing the evidence available at the time, the FDA agreed that diets rich in soy protein, as found in tofu and soy milk, for example, were capable of reducing blood levels of LDL, the "bad cholesterol." American products are now allowed to claim that daily consumption of twenty-five grams of soy protein, coupled with a diet low in saturated fat and cholesterol, *may* reduce the risk of heart disease. This claim is allowed as long as a serving of the food in question contains at least 6.25 grams of soy protein, less than 3 grams of fat, with no more than 1 gram of the saturated variety and less than 3 milligrams of cholesterol.

Not everyone was happy with the FDA's approval of a label claim, including some of the FDA's own scientists. Detractors maintained that excessive soy consumption can cause goiter, an enlargement of the thyroid gland, visible as a bulge in the neck. Genistein and daidzein, those famous soy isoflavones, were said to inhibit thyroid hormone synthesis by inactivating thyroid peroxidase, a vital enzyme. When thyroid hormone levels drop, the pituitary gland secretes more thyroid-stimulating hormone (TSH), causing the thyroid gland to grow and bulge. While there may be some laboratory evidence for the effect of isoflavones on the thyroid, there is no evidence that populations consuming a lot of soy show an increased risk of goiter or any other thyroid disease. Neither is there evidence, as has been suggested, that soy-fed babies show abnormal development or that high levels of phytoestrogens may cause feminization of boys. Three decades of soy formula use have revealed no correlation between consumption and developmental or hormonal abnormalities.

A more realistic concern is that the early studies about soy's benefits have not been corroborated by recent, better trials. Some twenty-two clinical investigations have been carried out since 1999, examining the effect of large amounts of soy protein on cholesterol levels. The results have been unimpressive: cholesterol was lowered on average by just 3 percent. Supplements containing isoflavones had no effect on cholesterol, and no clear evidence for cancer prevention has been found. Of course, this doesn't mean that soy-containing foods are not healthy choices. They are. But they are not a nutritional panacea. Replacing animal protein with vegetable protein is certainly helpful because it reduces the intake of saturated fats and cholesterol. Furthermore, soy contains a specific fatty acid, alpha-linolenic acid, which, as we have already seen, has been linked with a reduction in heart disease, independent of cholesterol lowering. A tofu salad still beats a smoked-meat sandwich when it comes to health—but unfortunately for most people, not when it comes to taste.

WHOLE GRAINS AND INSOLUBLE FIBER

Imagine I told you that a new dietary supplement shown to reduce the risk of heart disease, cancer, diabetes, and diverticulitis has just come on the market. And to boot, it even prevents weight gain. I suspect many of you would be off to the health food store, wallets at the ready. Alas, there is no such supplement. But there is a rather simple dietary modification that can lead to the benefits listed. Just eat at least three servings of whole grains every day! So how come people who would be ready to swallow pills to maintain their health are reticent about modifying their diet? Perhaps it is because most North American palates have become accustomed to the taste of breads, pastas, and cereals made from refined white flour. And we are creatures of habit. This habit, though, is worth breaking.

Simply stated, seeds are the plant structures capable of creating another plant, and those of the grass family are known as grains. Each grain is composed of three parts: the germ, the endosperm, and the bran. The germ is the component that can be fertilized by pollen, the endosperm is mostly starch and provides the energy needed for the germ to grow, and the bran is the tough fibrous outer coat that protects the seed. Grains can be cooked and eaten whole or milled to produce whole-grain flour. As our ancestors discovered, though, whole-grain flour doesn't have very good keeping qualities. Fats in the germ become rancid relatively quickly. On the other hand, flour made from the endosperm, which can be separated from the bran and the germ by a sifting process, keeps for a long time and has a more pleasing texture and taste. But if you want to make a nutritional comparison between whole-grain and refined flour—well, there is no comparison.

When most people think of whole grains, they immediately think of fiber, the part of the grain, found mostly in the bran, which defies digestion in the stomach and small bowel. But lack of digestion does not translate to lack of benefit. Dr. Dennis Burkitt, a British surgeon who worked as a medical missionary in Uganda, was the first to suggest a link between lack of fiber in the diet and disease, back in the 1960s. He noted that native Ugandans rarely suffered from colon cancer, heart disease, or diverticula (pouches) in their colons. British residents of Uganda, on the other hand, had a high incidence of these diseases. What was the difference? Diet. The British were accustomed to eating a low-fiber diet with a lot of refined white bread and meat, while the natives mostly feasted on plant foods rich in fiber. Eventually, there were theories aplenty to explain the protective effect of fiber; it absorbed or diluted carcinogens in the colon, it reduced transit time through the colon, it was digested by colonic bacteria to yield short-chain fatty acids, which had an anticancer effect. It also absorbed bile acids in the gut. These are synthesized in the liver to enhance digestion and are normally reabsorbed through the intestinal wall, but not when fiber is present. The liver then has to make more, and the raw material it uses to do this is cholesterol. The end result is that blood cholesterol level drops, and so does the risk of heart disease.

Since Burkitt's original observations, the links between consuming whole grains and health have been strengthened, although there have been some hiccups along the way, including the huge Nurses' Health Study that found no link between fiber intake and reduced risk of colon cancer. But this may be because even the high-fiber consumers were not consuming enough fiber. Most studies, though, have found a link. A large European survey of over half a million people in ten countries showed that the incidence of colorectal cancer could be reduced by some 40 percent if fiber intake is increased.

Then there is the Finnish paradox. Generally, in countries where heart attacks are common, so is colon cancer. Not so in Finland. The country ranks second in heart disease rate among industrialized nations, but a remarkable thirty-third in colon cancer incidence. The heart disease rate is explained by the very high fat content of the Finnish diet, but why the low incidence of colon cancer? It is probably due to lots and lots of fiber, mostly of the insoluble variety found in the whole-wheat bread the Finns are so fond of. This bread is great for the colon, but lowering cholesterol requires soluble fiber, such as that found in oats. It is also interesting to note that in spite of very high fat consumption, Finnish women have a low incidence of breast cancer. It seems that fiber reduces the levels of circulating estrogen that are linked with the disease. The Finns eat twenty-five to thirty grams (about one ounce) of fiber a day, which would be a challenge for most North Americans.

But there is more to whole grains than fiber. They compare with fruits and vegetables as a source of antioxidants, and contain various minerals and vitamins. Additionally, they provide lignans, which have established anticancer effects; rutin, which can reduce the risk of blood clots; and who knows how many other phytochemicals that may contribute to the benefits of a diet rich in whole grains. I could deluge you with numerous studies attesting to these benefits: studies about how three or more servings of whole grains can reduce insulin resistance, studies about how eating forty grams of whole grains a day significantly reduces middle-age weight gain, studies about whole grains lowering blood pressure, or ones about reducing cardiovascular risk by 30 percent with a couple of bowls of whole-grain cereal. But let's cut to the chase. How do you get three servings of whole grains a day? Easy. One serving is defined as thirty grams of hot or cold whole-grain cereal, one slice of whole-grain bread, or one-half cup (125 milliliters) of cooked whole-grain rice or pasta. Almost as easy as swallowing a pill, isn't it?

OATS AND SOLUBLE FIBER

I t would be interesting to take a look at Papa Bear's blood test. His triglycerides are probably high from slurping all that honey, but his cholesterol is likely to be just fine thanks to his love of porridge. In fact, the whole Bear family of the Goldilocks tale, with its penchant for oat porridge, can serve as a nutritional role model.

The Scots got this one right. Porridge is one of their staples. The oats are steeped not only in water and milk, but in a good dose of tradition as well. I understand the mush must be stirred clockwise with the right hand using a spurtle, which is sort of a scoopless wooden spoon, and it is to be eaten from a birchwood bowl. "Porridge sticks to the stomach and scrubs the bowels," the Scots maintain. True enough. Oats really do have a high satiety value. Essentially, this means they take a long time to digest and therefore keep you feeling full longer. Indeed, in one study comparing oatmeal with corn flakes for breakfast, researchers found that subjects who ate oatmeal consumed one-third fewer calories for lunch. Basically, oats can help you lose weight.

The "scrubbing the bowel" bit makes sense too. In more ways than one. Oats contain both soluble and insoluble fiber. Fiber is the structural part of plants, grains, fruits, and vegetables that cannot be broken down by enzymes in our digestive tract and therefore cannot provide nutrition. In other words, most of what you eat turns into you, but fiber just goes through. There are actually two kinds of fiber, insoluble and soluble. Cellulose is the classic insoluble fiber, whereas pectin, found in fruits, is an example of the soluble variety. The former keeps us regular, reduces the risk of diverticulitis, and helps eliminate substances that may play a role in colon cancer. But it is beta-glucan, the soluble fiber in oats, that is causing quite a stir. Solid research has shown that while oats produce no nutritional miracles (no single food does), consuming them regularly can lower blood cholesterol, reduce high blood pressure, keep our arteries healthy, and help control diabetes.

Some of this information about oats is not new. Just think back to the oat bran craze of a few years ago. Stores couldn't keep the stuff on the shelves. Rumors of a new shipment sent anxious shoppers rushing to the supermarket only to have their hopes dashed when they found that the booty had already been snapped up. Why was there such a feverish interest in a product that had traditionally been fed mostly to animals? Because some tantalizing studies showed that oat bran, the outer covering of the grain, was an excellent source of soluble fiber, which had the ability to reduce cholesterol. There was even a theory that explained how this happened. Beta-glucan absorbs water in the intestine and forms a viscous slurry that traps cholesterol from food as well as some of the bile acids needed for digestion. Since these compounds are made in the body from cholesterol, their removal from the digestive tract forces more to be synthesized. The result is a depletion of the cholesterol in the blood. But there was a problem. The public never got the proper message about how much oat bran had to be consumed to have an impact on blood cholesterol. This amount was not trivial.

To reduce blood cholesterol by roughly 5 percent, a person needs to eat three to four grams of beta-glucan a day. More is not better! At higher doses a feeling of fullness, a sense of bloating, and gas production become apparent. Now, a 5 percent reduction doesn't sound like a lot, but it can lower the risk of a heart attack by as much as 10 percent. This amount of beta-glucan is to be found in one cup (250 milliliters) of cooked oat bran or one and a half cups of oatmeal. Three packets of instant oatmeal will do it too. But oat bran cookies, oat bran chips and oat bran gum will not. Yet these silly products flooded stores hoping to capitalize on the oat bran mania. They had no effect on cholesterol and tasted lousy to boot. So it was little wonder that the oat bran fad quickly faded. Too bad. Because when consumed in the right amounts, oats can really deliver the goods. They can do more than just lower cholesterol; they can reduce blood pressure.

A pilot study in Minnesota focused on a group of patients who took at least one medication for hypertension. Half of them were asked to consume about five grams of soluble fiber per day in the form of one and a half cups of oatmeal and an Oat Square (an oat-based snack), while the other half ate cereal and snacks with little soluble fiber. Oat consumption reduced blood pressure significantly. Indeed, about 50 percent of the patients were able to give up their medication. How oats lower blood pressure is not clear, but it probably has to do with modifying insulin response. The pancreas secretes insulin, which is needed to allow cells to absorb glucose from the bloodstream after a meal. A glucose surge triggers a quick insulin response, but if such surges are frequent, insulin becomes less effective, and more and more needs to be produced. This leads to a condition known as insulin resistance. Researchers suspect that insulin resistance can play a significant role in elevating blood pressure by constricting blood vessels. Soluble fiber slows the absorption of nutrients from the gut and blunts the insulin response. This also explains why oats can help diabetics control their blood sugar levels.

And if that weren't enough to increase your appetite for oats, just consider that they contain a unique blend of antioxidants, including the avenanthramides, which prevent LDL cholesterol from being converted to the oxidized form that damages arteries. Given all of this, it isn't surprising that the first-ever health claim for a food allowed by the US Food and Drug Administration involved oat products. In 1997, the FDA ruled that food producers could claim that "soluble fiber from whole oats, oat bran or oat flour as part of a low-saturated-fat, low-cholesterol diet, may reduce the risk of heart disease." There was a caveat, though. The claim could be made only if a single serving of a food contained at least 0.75 grams of beta-glucan, no more than 3 grams of fat and no more than 1 gram of saturated fat.

Oats are not the only grain to contain beta-glucan. Barley is also rich in this soluble fiber. In fact, beta-glucan is found throughout the entire barley kernel as opposed to only in the outer bran layer, as it is in oats. Processing therefore does not remove the beta-glucan. This means that even refined products such as barley flour, barley flakes, or barley meal contain beta-glucan. So, justifiably, barley producers did not want to be left behind when it came to making label claims. They also petitioned the FDA and flooded the agency with studies attesting to the benefits of their product. Reviewers examined five clinical trials that had investigated the impact of consuming whole-grain barley and dry-milled barley products and concluded that there was a consistent lowering of blood cholesterol levels. Much to their joy, barley producers can now also claim that soluble fiber in barley in conjunction with a diet low in saturated fat and cholesterol can reduce the risk of heart disease. Bring on the oat bran for breakfast, sprinkled with flax and topped with berries of course, and the bean and barley soup for supper.

BEANS AND INOSITOL PENTAKISPHOSPHATE

A h, those beans. They can cause gas, and we don't need to carry out studies to prove it. But beans can also reduce the risk of heart disease and cancer. That claim, however, does require backing from scientific studies. Ideally, we would like to see what is called an *intervention study*, in which subjects are organized into two groups with virtually identical lifestyles except for one aspect of the diet. The experimental group, but not the control group, would be treated to a prescribed dose of beans. Both groups would then be followed for many years. Unfortunately, such intervention studies are very difficult to carry out and researchers are more likely to go for what are called *case-control* trials.

In such a trial, a set of patients suffering from a certain disease is compared with a roughly equal number of healthy people, matched for age, lifestyle, place of residence, physical activity level, smoking, body weight, and socioeconomic status. This is just what researchers at Harvard University did to try and tease out factors responsible for heart attacks in 2,118 individuals in Costa Rica. Much to their surprise, they found that eating a third of a cup of beans a day reduced the likelihood of suffering a heart attack by close to 40 percent! Just what in the beans is responsible isn't clear, but beans are rich in folic acid, magnesium, vitamin B_6, alpha-linolenic acid, and fiber, each of which in theory may have an effect on heart function.

Population studies are another way to gain insight into causes of disease. The health status of a large number of initially healthy subjects is continuously monitored, as are their lifestyles. Subjects periodically fill out food-frequency questionnaires, which are then analyzed in terms of specific dietary components. One of the best examples is the previously discussed Nurses' Health Study, which has followed thousands of nurses for many years, some of whom, as would be expected, developed breast cancer. Researchers theorized that the disease might be linked to a reduced intake of antioxidants, particularly flavonols. They therefore investigated the amounts of tea, onions, apples, broccoli, green pepper, and blueberries, all rich in flavonols, in the nurses' diets. The results were unexpected. There was no association between the total flavonol intake and breast cancer. But women who consumed beans or lentils twice a week were about 25 percent less likely to develop breast cancer than women who consumed them less than once a month. Just another example of how health effects are determined by the overall composition of a food, not by individual components.

Laboratory experiments and animal studies also offer clues about preventing and fighting disease. It may be these that will eventually shed light on why beans have anticancer properties. The secret just may lie in inositol pentakisphosphate, a substance found in beans, as well as in lentils, peas, wheat bran, and nuts. Tumor growth involves many chemical reactions, and specific enzymes play a major role in these. Phosphoinositide 3-kinases, first discovered in the 1980s, are involved in the development of lung, ovarian, and breast cancer. Substances that block the activity of these enzymes are therefore obvious targets of research. Most compounds that have shown efficacy have turned out to be too toxic for use, but researchers at University College in London have great hopes for inositol pentakisphosphate, which they have isolated from beans. This compound is remarkably nontoxic, even in large amounts. In laboratory studies on human cells, it inhibited angiogenesis, the process tumors use to grow the blood vessels they need to supply them with nourishment. But even more interesting results were found when human ovarian cancer cells were transplanted into mice.

Inositol pentakisphosphate had an effect comparable to that of cisplatin, the drug commonly used for ovarian cancer treatment. A further exciting finding was that this compound enhanced the effect of anticancer drugs.

In spite of the positive health effects of beans, people worry about including them in the diet. The fear is the potentially embarrassing emission of gases. Beans contain specific carbohydrates such as raffinose and stachyose, which are not broken down by our digestive enzymes in the small intestine. They therefore proceed to the colon, where they delight the resident bacteria that proceed to gobble them up. Unfortunately for us, these bacteria produce a number of gases as they dine on these carbohydrates, some of which, like hydrogen sulphide, are notoriously odiferous. But science may come to our rescue here. Marisela Granito and colleagues at Simón Bolívar University in Venezuela have been investigating this issue for years, and have now found that fermenting beans with two specific bacteria of the *Lactobacillus* species before cooking can lower the concentration of the offending carbohydrates by 90 percent without altering the nutritional value of the beans. They propose that the food industry can make use of these bacteria to market low-gas beans. Scientists in India have taken another approach. Using standard food irradiation technology, they exposed beans to gamma rays and found that this process, in combination with soaking the beans, eliminated most of the stachyose and raffinose.

Individuals react quite differently to beans in terms of how much gas they produce. Some can ingest copious amounts without a problem, others drive friends and family away after eating a single burrito. But even in the latter case, emissions can be decreased by slowly increasing bean intake. And in light of everything we know about the benefits of eating beans, it is worth making the effort. Replacing some of the meat in our diet with beans is a good idea. Maybe when Jack traded the family cow for the magic beans, he didn't strike such a bad bargain after all.

CABBAGE AND INDOLES

Like beans, cabbage doesn't have a great reputation. A British food critic once suggested that by comparison with boiled cabbage, "steamed coarse newsprint bought from bankrupt Finnish salvage dealers and heated over smoky oil stoves is an exquisite delicacy!" I have never tasted coarse newsprint, steamed or otherwise, but given the choice, I would go for the cabbage. I think we could all do with some more indole-3-carbinol.

The human body is a fantastic machine with a variety of defense mechanisms to protect itself against undesirable chemical intruders. A variety of enzymes is available either to convert these intruders into less harmful substances or to link up with them and eliminate them through the urine. These protective enzymes are cranked out by a cell's genetic machinery when receptors on the cell's surface are activated by the presence of potentially dangerous foreign substances. Way back in the 1950s, researchers noted that substances that caused cancer triggered the release of protective enzymes, but that unfortunately, in many cases, the enzymes were unable to eliminate the carcinogen completely. It was clear, though, that some test animals fared better than others. Apparently, they had more efficient enzyme-producing systems. There are human parallels here also. Not every smoker develops lung cancer. Why not? Do the lucky ones produce more protective enzymes? And if so, can we foster this trait?

A clue came when researchers noted that after rats were exposed to a carcinogen, they were more resistant to the effects of a second carcinogen. They appeared to be protected by the enzymes their cells synthesized in response to the first attacker. Obviously, exposure to a carcinogen is not a method we can use to protect ourselves against other carcinogens. But what if there were substances that had a chemical similarity to cancer-causing agents but were themselves not dangerous? Might they not trick cells into generating protective enzymes? By the 1960s it had become apparent that this was a real possibility. Chemicals in cabbage, as well as in other cruciferous vegetables (so-called because of the cross-shaped leaves) like broccoli, cauliflower, and Brussels sprouts were found to stimulate the production of protective enzymes. Soon researchers focused on one specific compound that had aroused interest because of its potential in the fight against breast cancer, namely indole-3-carbinol.

The connection here is through estrogen, the female hormone that has been linked with tumor promotion. The relationship between estrogen and breast cancer, admittedly, is not a simple one. Laboratory studies have shown that estrogen, like many chemicals in the body, undergoes a variety of reactions after it is produced. Its metabolism, as these reactions are collectively called, can take two alternative routes. One produces 16-hydroxyestrone, which seems to be the culprit in terms of stimulating the irregular multiplication of breast tissue cells. Alternatively, estrogen can be converted into 2-hydroxyestrone, a compound that is relatively inert. Both of these conversions are governed by specific enzymes, levels of which can be affected by various factors. This is where indole-3-carbinol comes in. It stimulates the protective enzymes that take estrogen down the safe path, meaning that there will be less exposure of breast tissue to nasty 16-hydroxyestrone molecules.

That's pretty interesting stuff, but it's also pretty abstract for most of us. Probably not enough to persuade people to rush into the kitchen and start boiling cabbage. But wait. Mice develop fewer mammary tumors when exposed to indole-3-carbinol. Rats exhibit less endometrial cancer. But things get even more interesting when we discover that researchers have actually fed 400-milligram capsules of indole-3-carbinol to women on a daily basis (roughly equivalent to the amount in half a head of cabbage) and determined that it really did affect the way that estrogen was metabolized. Within two weeks the levels of 2-hydroxyestrone, the good stuff, as it were, went way up. In fact, the levels rivalled those found in marathon runners, who are known to have a lower incidence of breast cancer.

So that's what happened to the pill poppers. But what about eating cabbage itself? Thanks to some Israeli research, we have an answer to that question as well. Eighty women on a kibbutz agreed to eat a diet high in cruciferous vegetables and submit their urine for analysis. The ratio of 2-hydroxyestrone to 16-hydroxyestrone in the urine increased, suggesting protection against breast cancer. It would be interesting to follow these women for a number of years and determine whether or not the breast cancer rate is actually lowered. There is a good chance that will happen, at least if we judge by some interesting epidemiological evidence available from Germany and Poland.

Breast cancer rates in the former East Germany were significantly lower than in West Germany, but after unification the disease pattern has become more equal. While obviously there were many differences in lifestyles between the two countries, it seems noteworthy that cabbage consumption was much higher in East Germany. This becomes even more meaningful in light of recent research carried out at the University of Illinois that examined why Polish women who have moved to the United States have a higher breast cancer rate than women in Poland. Cabbage is a staple in the Polish diet, but is less popular among Polish Americans. Was this a factor, the researchers wondered? So they stimulated test-tube colonies of human breast cancer cells with estrogen and added cabbage extract. The cabbage-treated cells grew more slowly. And it was not a question of using unrealistic amounts of cabbage extract; doses were those achievable by eating normal amounts of the vegetable. Furthermore, the experiments suggested that the effect was due not only to indole-3-carbinol. Other antiestrogenic compounds also seemed to be present in the cabbage juice.

You may now be ready to head toward the kitchen. Especially when you learn that cabbage is also high in vitamin K, which is receiving attention for its role in strengthening bones. The Nurses' Health Study found that those who consumed moderate to high amounts of vitamin K from vegetable sources had a 30 percent lower risk of hip fractures. Still need more convincing? Consider the fact that epidemiological studies have shown that there is a lower risk of colon cancer among people who claim to eat cabbage regularly.

There is a trick to cooking cabbage. Do not boil it in water! That's how you release the smelly sulphur compounds. The general rule with cabbage is that the more you cook it, the worse the smell. So just stir-fry the shredded cabbage in a little olive oil until it turns brown, and then cook it in its own steam for a few minutes. Add a little salt, pepper, and a touch of sugar. Then dump it on some freshly boiled thin noodles. You couldn't ask for anything better. Try it. It will taste a lot better than steamed Finnish newsprint.

BROCCOLI AND SULFORAPHANE

Paul Talalay eats his sprouts. Not only does he eat them, he sells them. He also sells tea made from them. But you won't find Talalay behind the counter in some health-food store. In fact, he scorns many of the overhyped, overpriced, under-researched products with which they entice customers. Where you will find the sprightly octogenarian is in the hallowed halls of Johns Hopkins University, where for many years he was the director of the School of Medicine's Department of Pharmacology and Experimental Therapeutics, and where he is now the John Jacob Abel Distinguished Service Professor of Pharmacology. Just mention Dr. Talalay's name in scientific circles and the conversation will immediately shift to "chemoprotection" and of all things, broccoli!

Talalay's fifty-year research career has focused on the prevention and treatment of cancer. As a young medical student he was intrigued by the case of a prostate cancer patient who responded dramatically to therapy with steroids. Were there other substances that could also affect this dreaded disease in a similar fashion? Maybe even prevent it? Talalay decided to devote his career to finding out. Finally, in 1992 he made a discovery that would not only tantalize the cancer research community, but would also splash his name across the pages of newspapers. Researchers had long known that populations eating lots of vegetables had lower rates of several types of cancer. But why? Was it some specific compound or set of compounds found in these foods that was responsible? Talalay seemed to have found an answer.

He had isolated from broccoli a compound called sulforaphane, which at least in laboratory experiments had decided anticancer properties. In mouse cells grown in tissue cultures, sulforaphane boosted the production of so-called phase II enzymes. These enzymes form part of the body's protection system against foreign intruders, including carcinogens. Glutathione-S-transferase, for example, binds to carcinogens and removes them from the body. Sulforaphane is regarded by the body as a foreign substance and cells gear up their biochemical machinery to produce phase II enzymes to eliminate it. The enzymes then remove sulforaphane, as well as many other foreign substances they encounter.

Inducing protective-enzyme formation in cell cultures is one thing; cancer protection in live animals is quite another. The obvious next step was to treat rats with sulforaphane before attempting to induce tumors in them with a known carcinogen. When dimethyl benzan-thracene, a potent inducer of breast tumors, was used, the results were astounding. Almost 70 percent of the control rats developed cancer while tumors were detected in only 35 percent of the rats that had been dosed with sulforaphane. Other experiments showed that sulforaphane also offered protection against colon cancer, a type of cancer that has been linked to carcinogens found in foods such as barbecued meat. But what did this mean for humans? After all, the rats' diet was not nearly as varied as a human's. Furthermore, the amount of sulforaphane that offered protection against cancer corresponded to the consumption of several pounds of broccoli a week.

Two possibilities now presented themselves. Either find a better source of dietary sulforaphane or investigate the use of isolated-sulforaphane supplements. The first option seemed more appealing because the nutritional literature is filled with examples of substances that perform quite differently when they are introduced in a pure form as opposed to being a component of food. Also, foods such as broccoli contain a number of other beneficial nutrients such as sele-nium, calcium, folic acid, and vitamin K. It was at this point that

Dr. Talalay learned that broccoli sprouts potentially could yield as much as fifty times more sulforaphane than adult broccoli. Why potentially? Because neither broccoli nor its sprouts actually contains sulforaphane; what they have is glucoraphanin, a compound that yields sulforaphane when it reacts with an enzyme, myrosinase. This enzyme is liberated when the plant's tissues are disturbed by chopping or chewing. Cooking destroys the enzyme, but fret not: bacteria present in our gut can also break down glucoraphanin to yield sulforaphane.

Talalay and his co-workers now studied various broccoli varieties and through a laborious process selected seeds with the highest content of glucoraphanin. So convinced were they of the potential nutritional benefits of the sprouts from these seeds that Talalay and plant physiologist Jed Fahey launched Brassica Protection Products, a company that would market "BroccoSprouts" with part of the profits going to research into cancer chemoprotection. These sprouts are guaranteed to yield twenty times as much sulforaphane as mature broccoli. To be sure, so far the benefits of sulforaphane have been shown only in cell cultures or in animals. Dr. Talalay would be the first to agree that reducing the risk of cancer takes more than just eating BroccoSprouts and that human trials are sorely needed. He has already begun to investigate whether phase II enzymes can be elevated in humans with BroccoSprouts, and he envisions trials in high-risk populations such as people with a family history of breast cancer or a history of colon polyps.

Commercializing broccoli sprouts led to another amazing discovery. Employees at the sprouting facilities took to snacking on the sprouts they were producing. A couple who had been suffering from stomach ulcers for a long time claimed that the sprouts had cured them! This was not a complete surprise because earlier experiments had shown that broccoli had some antibiotic properties, and the link between ulcers and infection with the bacterium *Helicobacter pylori* is well established. Test-tube studies quickly showed that purified sulforaphane killed forty-eight different strains of the bacterium. This is an exciting finding because *Helicobacter* infection is also a risk factor for stomach cancer. Preliminary studies have already shown that sulforaphane can reduce stomach tumors in mice—and at a dose that does not translate to a human having to eat truckloads. A daily snack of broccoli sprouts is all that is involved.

Pro-broccoli evidence is certainly piling up. But how should we best eat it? Raw broccoli is fine but most people prefer the cooked version, prompting the age-old question about potential nutritional losses due to cooking. A research paper published in the *Journal of Science of Food and Agriculture* in 2003 created quite a public stir with the finding that microwave cooking of broccoli resulted in an almost complete loss of antioxidant flavonoids.

Researchers described how they had cooked broccoli by boiling, steaming, or microwaving and then examined nutritional losses. Broccoli was chosen because of its reputation as a "healthy" vegetable, a reputation based on its containing sulforaphane as well as indole-3-carbinol, which we encountered in our cabbage discussion. Curiously, these were not the compounds the researchers monitored in the study. Instead they looked at various flavonoids, which are also supposedly beneficial because of their antioxidant properties. Surprisingly, microwave cooking resulted in a 97 percent loss of flavonoids as well as significant losses in other antioxidants, while steaming resulted in minimal losses. But the researchers were not the most adept cooks. First, they used too much water during microwaving, adding two thirds of a cup of water for one and a half stalks of broccoli, whereas the usual amount is just one or two tablespoons. Second, they cooked the vegetable longer than recommended; one to two minutes is sufficient. Both of these techniques could have resulted in the leaching of nutrients.

Microwaves work by heating water, and since water is distributed throughout broccoli, it is theoretically possible that nutrients are exposed to more heat during microwaving than during steaming, when heat has to travel from the surface of the florets to the interior.

But a study carried out in 2006 at the University of Essex in the United Kingdom has offered comforting results for microwave chefs. Instead of measuring flavonoid levels, the researchers actually determined levels of glucosinolates such as glucoraphanin. Cooking by steaming, microwaving, and stir-frying produced no significant losses, while boiling did show losses due to nutrients leaching into cooking water. So proper microwaving is fine!

Of course, what is really important is to make broccoli a regular part of the diet—raw, steamed, or microwaved. And we need to get rid of broccoli's reputation as the awful food that parents force their unwilling children to eat. George H. W. Bush certainly didn't help when he declared that his mother had forced him to eat broccoli, and now that he was president, he could finally avoid the vile vegetable. Well, the former president was still skydiving in his eighties, so perhaps being forced to eat broccoli as a child wasn't such a bad thing.

SPINACH, CORN, SQUASH, AND LUTEIN

Eating spinach, corn, or squash to see better? Sounds far-fetched, right? Let's look at the facts. First, though, we need a little primer on vision. It all starts when light enters the eye through a transparent dome-shaped covering called the cornea and then passes through another clear structure called the lens. Together, the cornea and the lens focus the light on the retina, which lines the back of the eye and transforms light into nerve impulses that can be registered by the brain as vision. Myopia, or shortsightedness, occurs when the image is focused in front of the retina, either because of an excessive curvature of the cornea or because the eye itself has an elongated shape. The macula is the central part of the retina and is responsible for controlling straight-ahead sight. If it doesn't function well, the center of the visual field becomes blurred. Such *macular degeneration* affects some 20 percent of the population over the age of sixty-five and often leads to significant visual impairment. What causes the macula to degenerate? The first clue came from a chemical analysis of the macula in the 1980s that revealed the presence of two pigments, lutein and zeaxanthin. People with healthy eyes had more of these substances in their macula than those who suffered from macular degeneration. Both lutein and zeaxanthin absorb light, particularly the blue wavelengths. These waves are the most energetic in the visible spectrum and are most likely to damage the macula after years of exposure. It seems that lutein and zeaxanthin may act like internal sunglasses, filtering out the potentially damaging rays. The actual damage occurs when light stimulates the production of free radicals in the eye, which then end up injuring the cells of the macula. Lutein and zeaxanthin not only filter blue light, they also can act as antioxidants, or free-radical scavengers. Zinc was also found in abundance in the macula. Its role in vision isn't clear but a number of enzymes require zinc for proper functioning.

Based on clues from the chemistry of the macula, a couple of potentially beneficial interventions for macular degeneration present themselves. We can try to increase the lutein and zeaxanthin content of the retina, or we can try to forestall damage by the use of antioxidants and zinc. In 1994, the National Eye Institute in the United States, part of the National Institutes of Health, decided to put the antioxidant-zinc combination to a test. More than 3,600 patients suffering from macular degeneration were enrolled and given various combinations of zinc and the antioxidants beta carotene, vitamin E, and vitamin C. One particular combination, a daily regimen of 500 milligrams of vitamin C, 400 IU of vitamin E, 15 milligrams of beta carotene, 80 milligrams of zinc, and 2 milligrams of copper (because zinc interferes with the absorption of copper, an essential nutrient) taken over six years, reduced the worsening of macular degeneration by 25 percent. No study so far has shown that taking supplements of any kind can prevent the disease. There are indications, however, that a diet high in lutein and zeaxanthin can.

A number of epidemiological studies have shown that a higher intake of foods rich in lutein and zeaxanthin—spinach, corn, and collard greens in particular—is associated with a lowered risk of macular degeneration. Intervention studies have corroborated the evidence. When Professor William Hammond at Arizona State University asked volunteers with healthy eyes to eat corn and spinach every day, their macular pigments increased significantly in as little as four weeks. In another study, fourteen people in the early stages of macular degeneration actually showed some improvement after starting to eat about five servings of spinach a week. Lutein and zeaxanthin supplements are available but they raise an issue. Too much of one carotenoid may interfere with the absorption of others. Lycopene in tomatoes, for example, may not be absorbed as effectively in the presence of a high dose of lutein. Lutein and zeaxanthin supplements need further investigation. The best bet is to get these carotenoids from the diet, so focus your eyes on green spinach, yellow corn, and orange squash. The greater the variety of colors in your shopping cart, the better you will see.

CURRY AND CURCUMIN

I t fights arthritis. It fights breast cancer. It fights prostate cancer. It fights colon cancer. It even fights Alzheimer's disease! Sounds like the usual hype for some dietary supplement scam, right? Actually, these claims are being made on behalf of a substance that is readily available in any grocery store. And the claims are not being made by hucksters but by reputable scientists, though they are careful to point out that, so far, most of the evidence comes from studies carried out on rodents, not humans. So what is this "hot" substance? Turmeric, the yellow spice used to add flavor to many dishes, curries in particular.

Turmeric is the ground-up root of an East Indian plant (*Curcuma longa*) belonging to the ginger family. It usually makes up 20 to 30 percent of curry powders, with other spices such as coriander, ginger, chili, black pepper, cumin, mustard, fennel, and cardamom added to the mix. But our focus here is not on turmeric's flavor; it is on its potential health effects, some of which were described thousands of years ago in India's ancient ayurvedic system of medicine. Turmeric was said to be good for stomach ailments, wound healing, and "blood cleansing." Today in India the spice is used as a household remedy for sprains and swellings, the same sorts of problems for which we would use aspirin or some other nonsteroidal, anti-inflammatory drug (NSAID). Could there be some ingredient in turmeric that has a similar function?

Modern research has identified curcumin, a compound that comprises about 10 percent of turmeric by weight, as the most likely candidate for health benefits. In laboratory studies curcumin inhibits the action of the cyclooxygenase-2 enzyme (COX-2), an enzyme that catalyzes the formation of pro-inflammatory prosta-glandins. And guess what other chemicals inhibit this enzyme? The nonsteroidal anti-inflammatories, including aspirin. It seems that these drugs and curcumin have something in common. But curcumin may even add another facet to the treatment of inflam-mation. In addition to its COX-2–inhibiting effect, turmeric also interferes with the production of a protein called NF-B, known to stimulate specific genes that code for inflammatory substances. Given all of this, we shouldn't be too surprised that researchers at the University of Arizona have shown turmeric can prevent joint inflammation in rats. A controlled trial with humans using stan-dardized doses of curcumin is sorely needed.

The use of aspirin and NSAIDs has been linked with a reduced risk of colon cancer, but the risks associated with taking these drugs, particularly the risk of gastric bleeding, preclude their use for pro-tection against the disease. Could curcumin offer safer protection? Possibly. Epidemiologists have noted that in India, where the popu-lation consumes an average of two to three grams of turmeric a day (containing 200 to 300 milligrams of curcumin), the incidence of colon cancer is roughly one-eighth of that in the West. And this may be more than a chance association, at least according to a small clini-cal trial at Johns Hopkins School of Medicine. Five patients who had a history of precancerous polyps in their colons were treated with 480 milligrams of curcumin and 20 milligrams of quercetin three times a day. Quercetin is an antioxidant found commonly in apples, onions, tea, and citrus fruits, and it has been associated with a re-duced risk of colon cancer. After six months of treatment, all five patients had fewer and smaller polyps. While the amount of quer-cetin used in the trial is readily available in the diet, the curcumin far exceeds the dose that could be provided by any curry. Still, this small study does lend support to the idea that when consumed regu-larly curcumin may be effective in preventing colon cancer.

Since Indians have one-quarter of our breast cancer rate and one-twentieth of our prostate cancer rate, researchers have cast a hopeful eye toward turmeric here as well. Dr. Bharat Aggarwal of the M. D. Anderson Cancer Center in Houston, perhaps the world's leading authority on turmeric, injected mice with human breast cancer cells taken from a patient whose disease had spread to the lungs. The mice developed tumors that were surgically removed to simulate a mastectomy. Some of the animals were then treated with curcumin, some with the widely used cancer drug paclitaxel (Taxol), some with a combination of the two, and some received no treatment. The most effective treatment was the combination, with only 22 percent of the animals developing lung cancer. Amazingly, curcumin alone was more effective than paclitaxel alone.

Researchers at Rutgers University have found similar results for prostate cancer induced in mice, this time studying the effects of curcumin and phenethyl isothiocyanate (PEITC), an anticancer compound in "cruciferous" vegetables such as broccoli, cauliflower, and cabbage. The mice were injected three times a week for four weeks, with the strongest tumor-retarding effects being found for the combination of curcumin and PEITC. Again, while it is hard to interpret such experiments in human terms, the study supports the regular consumption of the cruciferous vegetables along with turmeric.

Such a diet may even help prevent the buildup of amyloid plaque in the brain, a hallmark of Alzheimer's disease. Curcumin-fed rats produced less plaque after receiving beta amyloid injections in the brain than rats fed a normal diet. The curcumin-fed rats also outperformed other rats in maze-based memory tests. If you have trouble remembering all of this, maybe you could use a little help from turmeric.

At this point, there isn't enough known about the effects of turmeric to make a recommendation about consuming specific amounts, but adding turmeric-flavored vegetarian dishes to the diet is surely a good idea. Don't forget to add some pepper. It increases curcumin absorption a thousand fold. But be careful about how you eat. Turmeric can leave some nasty stains on fabrics. Don't panic, though. Rubbing stains with moistened detergent usually solves the problem. If not, 3 percent hydrogen peroxide will do the job.

CHOCOLATE AND FLAVANOLS

There is something unusual about the Kuna Indians living in the San Blas Islands of Panama. Or at least there was in the 1940s when a scientific paper described their extremely low blood pressure. The cause was unlikely to be genetic; Indians who had moved to the mainland did not have low blood pressure. Were they eating or drinking something on the islands that lowered their blood pressure? This is what interested Dr. Norman Hollenberg of Harvard Medical School. Examination of the Kuna lifestyle revealed that a beverage made from minimally processed cocoa beans was extremely popular. Could this be the key to the unusually low blood pressure of the natives?

Hollenberg knew that cocoa beans, like other natural products, were chemically complex. Researchers had isolated dozens of compounds from cocoa beans, as well as from chocolate made from cocoa beans. Some of these compounds had garnered attention in terms of health benefits, particularly a family known as the flavanols. Indeed, chocolate manufacturers had already been interested in flavanols, and the Mars Company was working on developing a tasty high-flavanol cocoa powder. This turned out to be a challenge because flavanols have an inherent bitter taste. In any case, when Dr. Hollenberg approached Mars, the company was happy to provide him with a supply of flavanols. It didn't take long before Hollenberg's studies showed that flavanols relaxed blood vessels and improved blood flow to the brain by 33 percent. Chalk one up for chocolate!

The blood-vessel-relaxation effect is not the only benefit that has been noted. At the University of California at Davis, Dr. Carl Keen has observed a flavanol-related "blood thinning" effect. It seems flavanols interfere with the activity of blood platelets, which make blood coagulate. The effect is similar to that of a daily baby aspirin, which people take to ward off heart attacks, many of which are caused by blood clots. There is yet another way that compounds in cocoa may help prevent heart attacks. At the University of Scranton, Dr. Joe Vinson examined the antioxidant effect of chocolate. Why look into this? Because one of the mechanisms by which coronary arteries get clogged involves the oxidation of low-density lipoproteins (LDL, the "bad cholesterol"). Presumably, if this oxidation can be curtailed, heart attack risk decreases. Vinson found, albeit only in the test tube, that cocoa powder and dark chocolate were very effective at reducing LDL oxidation. What does this mean in terms of how much chocolate people should eat? Not much, although a provocative preliminary study has found that about thirty-five grams · of defatted cocoa, roughly what is found in 1.5 liters, or seven cupfuls, of hot chocolate can have a significant impact on preventing LDL oxidation.

And the positive studies just keep coming. Dr. Roberto Corti at the University Hospital in Zurich showed that forty grams (1.5 ounces) of dark chocolate improved the flow of blood through the coronary arteries, whereas white chocolate, devoid of flavanols, had no effect. Dr. Jeffrey Blumberg at Tufts University randomly assigned twenty subjects to receive 100 grams of dark or white chocolate for fifteen days. The lucky subjects on the dark-chocolate diet saw their blood pressure and cholesterol drop and their response to insulin improve. Perhaps even more telling is a study carried out at the National Institute for Public Health and Environment in Holland. For fifteen years, researchers following the health status of 470 men, ages sixty-five to eighty-four, discovered that those who regularly ate cocoa products had lower blood pressure. But the really exciting finding was that the men who ate the highest amount of cocoa were less likely to die from heart disease. Still, this does not mean that people with high blood pressure, or indeed anyone else, should start guzzling chocolate. But if you are looking for a dessert, dark chocolate is a better choice than a doughnut.

A CocoaVia bar may be easier to justify than a chocolate-covered doughnut. This is the Mars Company's entry into the "functional food" market. Functional foods are those that aim to deliver more than just simple nutrition or taste, and they are now a $50 billion business in North America. Each CocoaVia bar contains 100 milligrams of flavanols. This means that two of these bars a day contain an amount of flavanols shown to have an effect on blood pressure and on platelet aggregation. Mars has even incorporated into each bar 1.5 grams of phytosterols, plant-derived compounds that can lower cholesterol levels. So far there have been no human trials to demonstrate the benefits (other than to the manufacturer) of consuming CocoaVia bars. But you never know where chocolate research will go. Dr. Hollenberg's work suggests that flavanols dilate blood vessels by triggering the release of nitric oxide, the same substance that is responsible for the activity of Viagra. Now if that effect stands up to clinical trials, women may be giving men chocolates on Valentine's Day.

Antioxidants such as flavanols are also thought to have an effect on the skin. Wilhelm Stahl and colleagues in Germany decided to put the matter to a scientific test. They had women consume a cup (250 milliliters) of either high- or low-flavanol cocoa daily for a period of twelve weeks. Women in the high-flavanol group showed reduced reddening of the skin upon exposure to ultraviolet light, increased skin thickness, better skin hydration, and a significant decrease in skin roughness and scaling. So chocolate seems to be good for our outsides as well as our insides. And if you are worried about chocolate causing acne, don't be. There is no scientific evidence for that common belief.

Gorging on chocolate while pregnant or lactating, however, may not be such a great idea, if we go by a report from the University of

Messina in Italy. Doctors found that a baby born to a mother who was a heavy consumer of cocoa and chocolate was irritable, jittery, and often cried inconsolably. All of the baby's symptoms resolved when the mother was told to give up chocolate—but one wonders whether she then became the crankiest person in the family.

COFFEE BEANS AND CAFFEINE

Just ponder this: If coffee were a synthetic concoction, it would not be allowed on the market! That's because coffee beans contain at least nineteen compounds that have been found to be animal carcinogens. As a matter of fact, the natural carcinogens that we ingest from coffee far outweigh the synthetic pesticide residues in our food supply that people fret so much about. Quite a disturbing thought, isn't it, considering many of us cannot start the day without a cup of joe. So how do we explain that there is no epidemic of cancer linked to coffee consumption? Simple. Amounts matter. The carcinogens in coffee are present in trivial amounts—much less than the amount that can trigger cancer in test animals. Moreover, coffee is a highly complex mixture, made up of over 2,000 compounds, including antioxidant polyphenols, with their reported anticancer properties.

Believe it or not, coffee is actually the number one source of antioxidants in the North American diet. Just ask University of Scranton chemistry professor Joe Vinson, whom we have already met because of his chocolate studies. Vinson has measured the antioxidant content of more than 100 foods and beverages and then coupled this information with frequency-of-consumption data to determine the top sources of antioxidants in our diet. Dates, for example, have the most antioxidants per serving, but let's face it, how many dates does the average North American eat per year? Precious few. But we sure drink a lot of coffee. So although in terms of antioxidants per serving coffee is outpaced by the likes of grapes and cranberries, it still contributes the most antioxidants to our diets because we consume so many more servings. Bananas, corn, and dried beans provide the most antioxidants after coffee. Again, bananas are not that rich in antioxidants, but North Americans, on average, eat about fifteen kilograms (thirty-three pounds) of bananas per person per year. That's about twice our apple consumption.

We're pretty safe in saying that coffee does not present a cancer risk. If it did, we would have seen the epidemiological evidence by now. But its role in hypertension and heart disease is a different story. Some studies have shown increased levels of hypertension-linked inflammatory molecules such as C-reactive protein and interleukin-6 in coffee drinkers, and at least one epidemiological investigation has found that Greeks consuming four or more cups of coffee a day suffer more frequently from high blood pressure.

Want to add to the coffee confusion? Then consider that cafestol and kahweol, both present in coffee beans, are known to stimulate the liver's production of cholesterol. They're commonly found in oil droplets released from coffee beans by the brewing process, but are retained by filter papers and are therefore are not found in filtered coffee. Other types of coffee, though, such as Scandinavian, Turkish, Greek, or "French press" could pose a problem because the liberated oils stay in the coffee. The Finns, for example, regularly drink seven to nine cups of boiled coffee a day, and they do have higher blood cholesterol levels. But then again, they also have a diet rich in animal fats.

Digesting all this data calls for a coffee break, and Dr. Wolfgang Winkelmayor of the Harvard School of Public Health may have provided one—at least for women. When he mined the Nurses' Health Study for data, he found no relationship between coffee consumption and blood pressure. In fact, women who drank the most coffee seemed to develop some protection against hypertension. While that finding is still under investigation, coffee, at least until the next study comes along, has been cleared of causing high blood pressure. As far as heart disease goes, we can refer to one of the largest and best epidemiological studies ever conducted, the Health Professionals Follow-Up Study, overseen by the Harvard School of Public Health. This study followed over 45,000 men for several years and found that total coffee intake was not associated with coronary heart disease or stroke even when the men drank more than four cups a day.

Despite this, some worries crop up about coffee. Over three cups a day may increase rheumatoid arthritis symptoms. Connections to osteoporosis, birth defects, or fibrocystic breast disease have also been mentioned. While no clear links have been found, most health authorities recommend that pregnant and breastfeeding women limit coffee consumption to two cups a day. There is no doubt that coffee increases urinary frequency, a feature that men with prostate problems have to consider.

Enough about coffee worries. Let's talk about some more positives. Several studies have demonstrated that four to five cups a day can reduce the risk of type 2 diabetes by as much as 30 percent. The theory is that caffeine, along with chlorogenic acid and compounds called quinides—all present in coffee—increase energy expenditure and lead to weight loss. In addition, chlorogenic acid seems to keep sugar from being absorbed from the gut into the bloodstream. Now for the really interesting stuff. Coffee may even aid in the war against Parkinson's disease! This tragic degenerative disease is caused by a deficiency in dopamine, a neurotransmitter, which nerve cells use to communicate with each other. One cause of such deficiency is thought to be overactivity by adenosine, another neurotransmitter. And guess what? Coffee reduces adenosine activity. Interestingly, adenosine also has a soporific effect, which may explain why drinking coffee allows us to work that extra shift and helps students to pull all-nighters.

It is fair to conclude that moderate coffee consumption poses no risk and may even have some benefits. Better, though, to leave out the sugar and the cream, which pack lots of extra calories. And let's remember that for many people coffee is a great source of pleasure, one we might not have learned about if it was not for Kaldi's goat.

According to an oft-quoted legend, some 1,200 years ago, Kaldi, a Yemeni goatherd, found one of his animals in a highly excited state, darting back and forth and bleating frantically. He discovered that the goat had become bewitched by snacking on some strange violet berries. Confused and bewildered, Kaldi ran to his imam for help. The sage spiritual leader turned out to have a scientific spirit and made a brew of the berries. Upon sampling the concoction, he felt a rush of energy and alertness. And it was thus that the effect of caffeine was discovered, and the name *kahveh*, meaning "invigorating" in Arabic, given to the juice made from the odd little berries.

Some people want the extract without the caffeine and its side effects. They like the taste of coffee but not the jitteriness caffeine can produce, so they choose decaffeinated coffee. Several processes can be used to remove most of the caffeine from coffee. They all rely on the fact that caffeine is soluble and they all start by soaking the coffee beans in hot water. This process extracts the caffeine but it also extracts many of the flavor compounds. The idea is to remove the caffeine from this extract and reintroduce the flavor components back into the beans. First, you need a solvent that does not mix with water, and in which the caffeine is more soluble than it is in water. The classic ones used have been methylene chloride and ethyl acetate. Since ethyl acetate is found in some fruits and vegetables, it is often described as a "natural" substance. This is a crock, because ethyl acetate is not found naturally in the amounts used in the decaffeination process. In any case, the water extract is shaken with the solvent, which dissolves the caffeine. Since the solvent does not mix with water, it can be readily separated. The beans are then resoaked in the water to reabsorb their flavor. Of course, not all the flavor compounds are reabsorbed, so decaf will never taste exactly like regular coffee. Note that the extracting solvent never comes into contact with the beans themselves, so there is essentially no residue of the solvent in the coffee. Nevertheless, people have been concerned about the use of chemicals to decaffeinate their coffee, and processors have had to come up with other systems.

Highly compressed carbon dioxide gas can be used to extract the caffeine from the beans. This is an efficient process, and it leaves no residue to worry about. The Swiss water process is also heavily promoted. After the beans are soaked in hot water, the water is passed through activated carbon filters that absorb the caffeine but not the previously dissolved flavor compounds. A fresh batch of coffee beans, which contains caffeine, is then soaked in this "decaffeinated" water. Since the water is already saturated with the flavor compounds, more will not dissolve out of the beans. But since there is no caffeine in the water, the caffeine from the beans will dissolve into the water. Since only water is used in this process, there is no worry about any solvent contamination.

And how about instant coffee? Nobody will claim that it challenges the flavor of a good filter brew or an espresso, but it is convenient— no fooling around with grinders or coffee machines, no mess, no grounds to get rid of. Just add hot water and drink. But what exactly is that powder in the jar? The answer lies in the mountains of Guatemala. It was there in 1906 that an American engineer named George Constant Louis Washington brewed a regular pot of coffee. He must not have been paying much attention because the pot boiled over, spewing coffee everywhere. By the time Washington remembered the pot, the boiled-over coffee had dried around the spout into a powdery brown residue. On a whim, Washington tasted the powder and was pleasantly surprised. And he was really tickled when he added the powder to some hot water and found it dissolved to produce an acceptable cup of coffee.

Washington, of course, had not set out to invent instant coffee, but others before him had tackled the problem. The general idea was to evaporate the water from brewed coffee and attempt to reconstitute the residue into an acceptable beverage by adding water. Results were terrible. The reconstituted coffee had a burned taste. That's because these attempts were made at sea level where the boiling point of water is of course 212°F (100°C), and heating coffee at that temperature produces a variety of bitter compounds.

Washington's luck was that the mountains of Guatemala are high, and that the boiling point of water decreases with altitude—which is why it takes a much longer time to make a hard-boiled egg atop Mount Everest than at sea level.

Washington's coffee pot boiled over at about 185°F (85°C), and at that temperature far fewer bitter compounds are produced. Being an engineer, Washington figured out what had happened and opened up the George Washington Coffee Refining Company in Brooklyn in 1909. Here he produced the first batches of commercial instant coffee by "low-temperature boiling under reduced pressure." American soldiers during the First World War welcomed instant coffee in their battle rations. Today, instant coffee production has been refined, but the basic idea is still to evaporate the water at low temperatures. This can be done by heating the coffee under a vacuum or by squirting the coffee under pressure through tiny holes to produce a fine spray that dries almost instantly as it meets a blast of hot air. There is also the freeze-drying method wherein the coffee is frozen and then placed in a vacuum chamber. The water is pumped off, going directly from the solid phase into the gaseous one. This technique probably delivers the best flavor.

After all this coffee brewing, what should you do with the leftover grounds? Keep them, just in case an elephant tramples over your lawn. Apparently coffee grounds are excellent for removing the smell of elephant urine.

GRAPES AND RESVERATROL

They feast on croissants that ooze butter. They eat creamy cheeses and fat-filled pastries. Breakfast is *pain au chocolat*, washed down with espresso. There is no oatmeal in sight. I suspect most have never heard of flaxseed. Yet the French have the lowest death rate from heart disease in the European Union, and when we compare this rate with North America's—well, there is no comparison. Our incidence of heart disease is double that of the French, who are also much slimmer than Canadians and Americans. How do we explain this situation, which has been dubbed the "French Paradox"? According to some researchers the secret is to be found in wine, particularly red wine. More specifically, they point a finger at resveratrol, an antioxidant compound in the polyphenol family.

The simplified argument goes like this. Most heart attacks occur when a blood clot forms in a coronary artery and chokes off the flow of blood, starving the heart of oxygen. Blood clots form when the endothelium, the inner lining of the artery, is damaged. Such damage is associated with the formation of deposits called plaque, which in turn are linked to the presence of excessive amounts of cholesterol in the blood. But cholesterol carries out its dirty work only when it undergoes a chemical change stimulated by the presence of oxidizing agents such as free radicals. Oxidized cholesterol, then, is the real culprit, and if its production can be curtailed, the risk of a heart attack can be reduced. Antioxidants can do this—at least in the test tube.

Resveratrol, as it turns out, is not only an effective antioxidant, it can also reduce the blood's clotting ability. Little wonder then that resveratrol pills have begun to appear in health food stores. The efficacy of these pills, however, is highly questionable, since isolated resveratrol is an unstable compound. Special care has to be taken to preserve it, for example, by packing it in airtight capsules under a nitrogen atmosphere. Such products do exist and have been shown to have antioxidant effects on human cells in cultures, but there is no evidence that they do anything in live animals, never mind in humans.

While I find the resveratrol research engaging, so far it hasn't convinced me to up my intake of red wine. The truth is that I'm just as happy to have a glass of water with my dinner, and it doesn't even have to be bottled water. But I just may have to rethink my beverage preference in light of some interesting research coming out of Harvard Medical School. Although it doesn't exactly relate to the French Paradox, it is still pertinent. Why? Because we would all like to live longer. Molecular biologist Dr. David Sinclair and his colleagues have found a way to increase lifespan—at least for yeasts—by feeding them red wine! All right, so yeasts aren't people—or even rodents. But what works for yeasts may work for humans, because it seems that we also have a version of the gene that allows yeasts to live longer when exposed to red wine.

Yeasts are excellent organisms to use to study aging because they are easy to work with in the laboratory and have relatively short life cycles. As early as 1991, researchers had discovered that some yeasts lived longer than others. Why, was the big question. That was answered by Dr. Leonard Guarente of the Massachusetts Institute of Technology, who found that the long-lived yeasts produced an enzyme called sirtuin, which had the ability to repair damaged DNA. Strangely, the gene that codes for this enzyme, termed SIR2 ("silent information regulator"), becomes more active when yeast cells are starved of nutrients. This is not totally surprising because evidence exists that not only yeasts, but also fruit flies, rodents, and monkeys all live longer when put on a calorie-restricted diet. This characteristic is probably an evolutionary vestige: when food is in short supply, reproduction is difficult and organisms need to live longer so as to postpone breeding until conditions improve. Some research has shown that humans who eat roughly 30 percent fewer calories than generally recommended live longer than average.

Researchers' attention turned to possible ways to activate the gene that seems to code for the enzyme that plays a role in increased life span. They started systematically to examine chemicals that could possibly increase enzyme activity. It didn't take long to find one that aroused their interest. Resveratrol performed remarkably well, mimicking the effect of calorie restriction. And let's face it, drinking a glass of red wine every day is a lot more pleasant than reducing calorie consumption by 30 percent. According to the research, one glass (four ounces) is all that is needed to increase life expectancy by ten years, if indeed the effect on humans is similar to that on yeasts. There seems to be a sort of justice in this research. Yeasts convert grape juice into wine, and wine repays the favor by providing resveratrol to allow the yeasts to live longer.

When Dr. Sinclair progressed from yeasts to mice, he found an interesting result, one that certainly captured the imagination of journalists around the world. "Red wine substance appears to counter bad health in fat mice," screamed the headlines. Dr. Sinclair fed one group of mice a standard laboratory diet, another group an unhealthy diet with 60 percent of the calories coming from fat, and a third group the same unhealthy diet supplemented with regular doses of resveratrol. As expected, the mice in the second group became obese, showed signs of diabetes and heart disease, and died prematurely. The mice in the resveratrol group also became fat, but they remained healthy and lived as long as the animals that ate a normal diet and stayed thin. Before you reach for the corkscrew, note that the amount of resveratrol given the mice was roughly equivalent to that found in 100 bottles of red wine. By all means, though, if you have obese mice and want them to live a long time, feed them resveratrol supplements.

There is also some intriguing preliminary evidence that drinking red wine may prevent Alzheimer's disease. "Preliminary" is the key word, but let's face it, all significant findings start out with preliminary research. Dr. Jun Wang at New York's Mount Sinai School of Medicine worked with mice that had been specially bred to produce high levels of a protein called beta-amyloid. This protein can accumulate in the brain and has been implicated in Alzheimer's disease. When Dr. Wang put such mice on a diet that included an amount of red wine equivalent to a couple of glasses a day for a human, he found something amazing. The mice were better able to solve mazes than a control group of animals that had consumed alcohol equivalent to the alcohol content of wine. After the experiment, the brains of the mice were examined, and those in the wine group had significantly fewer deposits of beta-amyloid. Furthermore, Dr. Wang doused beta-amyloid protein with red wine in a test tube and discovered that the structure of the protein was altered in a fashion that prevented it from being deposited in the brain.

Research into resveratrol is clearly promising, but so far there is insufficient evidence to recommend that people who normally do not drink red wine take up the practice. And there are risks. Not much more than a couple of glasses a day has been associated with breast and oral cancers, and there are the social consequences of increased alcohol intake.

Returning to the French Paradox, the answer to why the French are slimmer and are less likely to have heart disease may lie not in what they drink, but in what they eat—or rather what they *don't* eat. The French simply eat fewer calories than the majority of North Americans, and their obesity rate is only about 7 percent compared with about 33 percent for Americans.

In 2003, Dr. Paul Rozin of the University of Pennsylvania and his associates compared portion sizes in France and the United States, weighing servings in eleven comparable pairs of eateries in Paris and Philadelphia. These ran the gamut of pizzerias, fast-food outlets, and ethnic restaurants. The average portion size in the Paris restaurants was 277 grams as compared with 346 grams in Philly—a 25 percent difference. The American Chinese meals were a stunning 72 percent heftier than those served in the Parisian Chinese restaurants. Rozin also found that portions of packaged foods were larger in the United States. An American candy bar was 41 percent larger, a hot dog was 63 percent bigger, and even single yogurt servings were much larger.

Then there was another finding. The French don't wolf down their meals, they take their time. Even at fast-food joints like McDonald's, they take longer to eat their burgers and fries. Americans spend fourteen minutes "enjoying" their fast food while the French linger for some twenty-two minutes. The French also don't eat at their desks and they don't eat on the run. In total, an average American spends an hour a day eating while a French person eats for some 100 minutes. It seems the French eat less and enjoy it more.

French wine producers prefer to credit red wine's antioxidants for producing the French Paradox, and they have produced a white wine with similar properties. A team of wine researchers at Montpelier University have come up with a chardonnay called "Paradoxe Blanc" that has almost the same antioxidant potential as red wine. They found that if the grapes were macerated with the skins and seeds and the fermentation temperature increased, the polyphenol content of the wine increased dramatically.

Furthermore, these scientists managed to show that the chardonnay really has an effect on the antioxidant capacity of the blood. They destroyed some of the insulin-producing cells in the pancreas of rats to make the animals diabetic, because diabetes is known to reduce the antioxidant capacity of the blood. Then they administered the new chardonnay to the critters for six weeks and found that the antioxidant capacity was restored. So those drinkers who prefer white over red should track down some Paradoxe Blanc. Of course, the real paradox is why people just don't eat more fruits and vegetables, which have more antioxidants than red or white wine!

While the role of red wine in the French Paradox may be ambiguous, this alleged connection has spawned some other possibly fruitful lines of research. Dr. Joseph Anderson of the State University of New York at Stony Brook spends much of his time looking through a colonoscope searching for cancers and precancerous polyps in people's colons. Because alcohol consumption has been suspected as a contributing factor to colorectal cancer, Anderson decided to survey his patients about their alcohol habits. He found that beer or spirit consumers who drank more than one drink a day were significantly more prone to colorectal tumors than moderate drinkers or abstainers. Red wine drinkers, on the other hand, seemed to be protected from the disease. Only 3 percent of those who drank at least three glasses of red wine a week had either cancerous or precancerous lesions, as compared with 10 percent of those who drank no alcohol. White wine showed no benefit. Anderson thinks that resveratrol, which is found far more extensively in red grapes than in white, is responsible.

There appears to be some theoretical justification for this possibility. Prostaglandins are compounds produced in the body that serve a multitude of functions, but some can suppress immunity and even stimulate tumor cell growth. Resveratrol has been shown to block an enzyme, cyclooxygenase-2, which catalyzes the conversion of arachidonic acid (a dietary component) into the problematic prostaglandin. In separate experiments, resveratrol has been shown to be a potent scavenger of potentially harmful free radicals. Still, the resveratrol connection may be overly simplistic, given that there are many other polyphenols in red wine that may contribute to the overall antioxidant effect.

Dr. Janet Stanford of the Fred Hutchinson Cancer Research Center in Seattle shares the view that resveratrol may be the key component. She studied alcohol consumption in 750 men with recently diagnosed prostate cancer and in a similar group of healthy men. Drinking at least four glasses of red wine a week was associated with a 50 percent lower risk. Stanford hypothesizes that resveratrol's ability to rid the body of free radicals, its anti-inflammatory effect, and its tendency to hold down cell growth all play a part in its protective role.

Since free radicals have also been implicated in the neurological damage that follows a stroke, Dr. Sylvain Doré and colleagues at Johns Hopkins University investigated resveratrol's potential to prevent such damage. Oral pretreatment of mice with resveratrol resulted in a 40 percent decrease in the area of the brain damaged by the induced stroke. Doré even managed to tease out the specific mechanism involved in the protection, namely an increased level of heme oxygenase, an enzyme known to shield nerve cells against free-radical damage. Based on his mice experiments, Doré thinks that a couple of glasses of red wine a day could produce a prophylactic effect against stroke damage in humans. But that's just a guess—just like almost everything else about red wine.

Now let's get back to the French Paradox. Actually, there may not even be one. Some researchers argue that the French use different criteria in ascribing causes of death and that some cases that would be described in North America as "cardiac" would not necessarily be described this way in France. In any case, while the extent of a reduced risk of heart disease in France is debatable, there is one thing we do know from reliable statistics: the French life expectancy is roughly the same as it is in North America. They don't live any longer; they just exit by a different route.

WHEAT AND GLUTEN

Just ask people what they worry about most in their food supply and they'll round up the usual suspects. Their thoughts will drift to nitrites, sulphites, food-coloring agents, artificial sweeteners, monosodium glutamate, or genetically modified organisms. Well, think again. We are far more likely to be harmed by a commonly occurring natural component in food than any of these. Gluten, a protein found in wheat, barley, rye, and to some extent oats, can provoke health problems in a significant percentage of the population. Celiac disease, as gluten intolerance is usually called, may be much more common than we think.

Dr. Samuel Gee of Britain was the first to provide a clinical description of the disease in 1888. He painted a disturbing picture of young children with bloated stomachs, chronic diarrhea, and stunted growth. Dr. Gee thought that the condition might have a dietary connection and put his young patients, for some strange reason, on a regimen of oyster juice, which, not surprisingly, proved to be useless. Willem K. Dicke, a Dutch physician, finally got on the right track when he made an astute observation during the Second World War. The German army had tried to starve the Dutch into submission by blocking shipments of food, including wheat, to Holland. Potatoes and locally grown vegetables became staples, even among hospitalized patients. Dicke noted that his celiac patients improved dramatically! Moreover, in the absence of wheat and grain flours, no new cases of celiac disease were seen.

By 1950 he had figured out what was going on. Gluten, a water-insoluble protein found in wheat, was the problem. As later research showed, celiac patients' immune systems mistake a particular component of gluten, namely gliadin, for a dangerous invader and mount an antibody attack against it. This triggers the release of molecules called cytokines, which in turn wreak havoc with the tiny fingerlike projections, the villi, that line the surface of the small intestine. The villi are critical in providing the large surface area needed for the absorption of nutrients from the intestine into the bloodstream.

In celiac disease, the villi become inflamed and markedly shortened, effectively reducing their rate of nutrient absorption. This has several consequences. Unabsorbed food components have to be eliminated, which often results in diarrhea. Bloating can also occur when bacteria in the gut metabolize some of these components and produce gas. But the greatest worry is loss of nutrients. Protein, fat, iron, calcium, and vitamin absorption can drop dramatically and result in weight loss and a plethora of complications. Luckily, if the disease is recognized, and a gluten-free diet is followed, patients can lead a normal life.

Diagnosis of celiac disease involves taking a biopsy sample from the duodenum, the uppermost section of the small intestine, via a gastroscope passed down through the mouth. Microscopic analysis shows the damaged villi. Recently, blood tests have also become available. A commonly available one tests for the presence of antigliadin antibodies but it is not foolproof. Only about half the patients with positive results actually show damaged villi upon biopsy. The anti-tissue transglutaminase test (anti-tTG) is a much better diagnostic tool, but is available only in specialized labs.

There is a great deal of interest in these tests because of their potential value in identifying celiac cases and perhaps even in screening the population. Celiac disease, which has a genetic component, does not necessarily begin immediately after gluten is first introduced into the diet. The onset of disease can occur at any age. In adults, the symptoms are usually much less dramatic than in young children. The first signs often are unexplained weight loss and anemia due to poor iron and folic acid absorption. Stools tend to be light in color, smelly, and bulky because of unabsorbed fat. Symptoms can include a blister-like rash, joint and bone pain, stomach ache, tingling sensations, and even headaches and dizziness. Identification of celiac patients is important not only because much of the misery can be prevented by a gluten-free diet, but also because a recent study showed that over a thirty-year period the death rate among celiac patients was double that expected in the general population. The risk rose with increasing delay in diagnosis and poor compliance with diet. The major cause of death was non-Hodgkin's lymphoma, a type of cancer known to be associated with celiac disease. A less severe but more common complication than cancer is osteoporosis, resulting from poor absorption of calcium and vitamin D.

Unfortunately, a gluten-free diet is not that easy to follow. Wheat and barley crop up in a wide assortment of products. Patients have to become veritable sleuths and learn that foods as diverse as ice cream, luncheon meats, ketchup, chocolate, and even communion wafers can contain gluten. Luckily the Celiac Association has excellent information on dietary do's and don'ts, and a large assortment of gluten-free products based on rice, corn, and soy is now commercially available, including communion wafers.

The plan of action for biopsy-diagnosed celiacs is clear. They must adhere religiously to a gluten-free diet to eliminate symptoms and reduce the risk of osteoporosis and lymphoma. But what about people who have no overt symptoms but show a positive blood test? Surveys indicate that one in about 200 people may fall into this category. Their biopsies may show normal villi, but these people are considered to have latent celiac disease that may become symptomatic years later. Others may have flat villi but no symptoms, and they are judged to have silent celiac disease, which can become aggressive at any time. Should they be put on a preventative, difficult-to-maintain diet? At this point nobody really knows, since much still remains to be learned about the effects of gluten. Recently, for example, researchers discovered that celiac patients who complained of headaches showed brain inflammation on MRI scans and that the problem resolved on a gluten-free diet. Some individuals have provided anecdotal and controversial evidence that the condition of some autistic children improves when gluten is eliminated from their diet. However, there is no evidence that they have celiac disease.

So it certainly seems that we have not yet uncovered all of gluten's potential for mischief. On the other hand, one intriguing possibility to reduce exposure to gluten has emerged. Preliminary research suggests that it may be possible to remove gluten's offensive component by genetically modifying wheat. That would be a boon to celiac patients and perhaps even to those of us who may be suffering in silence.

CINNAMON AND METHYLHYDROXYCHALCONE

Apple pie to reduce blood sugar? Sounds far-fetched, but giving apple pie to a group of diabetics did trigger some new insight into diabetes and even served up a possible treatment. No, the apples weren't the key to reducing blood glucose; it was the cinnamon flavoring! Richard Anderson at the Human Nutrition Research Center in Beltsville, Maryland, was interested in the effect of various foods on type 2 diabetes and gave some subjects servings of apple pie, expecting their blood glucose to soar. But it didn't. Instead, the pie seemed to actually lower blood sugar levels. Anderson knew that it was unlikely that any of the major ingredients in the pie was responsible for this unexpected effect, but cinnamon was a possibility. After all, numerous folkloric remedies had long associated cinnamon with relief from all kinds of ailments.

Anderson decided to put this surprising discovery to the test and enlisted sixty type-2 diabetics for a study. Subjects were given small doses of cinnamon, ranging from as little as a quarter teaspoon (roughly one gram) to less than two teaspoons a day for forty days. A control group was given capsules with wheat flour, a substance that does not affect blood sugar levels. The results were sweet indeed. Not only did cinnamon reduce blood sugar levels, in some cases by as much as 30 percent, but it also lowered LDL (the "bad cholesterol") and triglyceride (fat in the blood) levels as well! Even twenty days after the cinnamon study had ended, blood glucose levels stayed low, suggesting that cinnamon does not have to be consumed every day in order to produce a notable effect in the body. Surprisingly, subjects who consumed only a quarter teaspoon of cinnamon did as well as those who took higher doses. Researchers at Columbia University duplicated these results. In a placebo-controlled, randomized, double-blinded study, they found that a gram of cinnamon a day lowered fasting blood glucose by 17 percent after eight weeks. Impressive!

Diabetes, characterized by a higher than normal blood sugar level, is a serious disease. It can cause kidney and cardiovascular problems, eye damage, and impaired circulation. Two primary varieties of the disease exist. Type 1 usually presents at a young age and is due to the pancreas not producing enough insulin. Type 2, especially common in overweight people, commonly manifests itself in adulthood. While the pancreas still produces insulin, the hormone cannot do its job properly because the body's fat, muscle, and liver cells have become resistant to it. The job of insulin is to serve as a gatekeeper for the entry of glucose into these cells; if its work is impaired, glucose isn't absorbed and it accumulates in the blood, eventually causing damage.

While type 1 diabetes has to be treated with insulin injections, type 2 can often be controlled by paying attention to diet. The link between type 2 diabetes and obesity, especially abdominal obesity, is strong. Fat cells secrete adipokines, a group of hormones that impair glucose tolerance, and abdominal fat seems to be the most hormonally active. Moreover, the body becomes insensitive to insulin's regulatory effect because of long-term sugar and insulin overload. It is as if the body went on strike after being overworked.

With the increase in obesity rates, it isn't surprising that type 2 diabetes is becoming an epidemic in North America, even among children. While antidiabetic drugs are effective, many people are bent on exploring alternative therapies, including food supplements and herbal products. Possibly, when used in a complementary fashion, conventional medicine and alternative therapies can have a synergistic effect. Indeed, researchers have been examining various dietary substances for their potential blood glucose–lowering properties. Green tea, but not black tea, holds some promise, and so does coffee. The catch with coffee, though, is that it takes six cups a day to have an impact on blood sugar, and that amount of coffee comes with a load of caffeine. Fortunately, the active ingredient responsible for coffee's blood glucose–lowering effect is not caffeine, but chlorogenic acid, a compound that can be isolated and perhaps formulated into pills.

If coffee and tea do not appeal, red wine is another option, probably due to its resveratrol content. However, just as with coffee, the effects are seen only when consumption is more than moderate. Three glasses a day will do it, but that amount of alcohol can increase the risk of some cancers. Claims have also been made that other plant-derived materials reduce blood sugar levels, but with rather weak evidence. Fenugreek, bitter melon, Korean ginseng, gymnema (an herb taken from a vine that grows in India), onions, and flaxseed are just some that have been investigated. Interpreting the studies, however, tends to be problematic. Take ginseng, for example (or don't take it, depending on the study).

Andrew Scholey and his team from Northumbria University in the United Kingdom found that G115, a commercially prepared ginseng extract, significantly lowered blood glucose levels, but only in healthy, fasting individuals. Conversely, people who were administered ginseng together with a glucose drink had a greater rise in blood glucose than would be expected from glucose alone. The implication is that it is better for diabetics to leave ginseng products alone, especially since there are many different species of ginseng, all with different physiological effects, and commercially available preparations are not always pure or standardized.

Now back to cinnamon. To reduce blood glucose in type 2 diabetes you don't have to down a truckload of the spice. It appears that one gram a day, or roughly one-quarter of a teaspoon, is the optimal amount. (Type 1 diabetes appears not to respond to cinnamon.) Of course, as with any other intervention, we have to raise the question of possible harm. Cinnamon contains coumarin, a naturally occurring compound that can cause liver and kidney damage if consumed in high doses. The amount of coumarin depends on the specific cinnamon species. Ceylon cinnamon, also known as "true cinnamon," contains much less coumarin than Cassia cinnamon, the version usually sold as a powder in North America. Cinnamon powders are indistinguishable, but the "sticks" from which they are made can be readily identified. Ceylon cinnamon sticks are made of many thin layers and are easily ground into a powder while Cassia sticks are composed of one thick, hard layer. To circumvent coumarin contamination, some companies have launched preparations made by extracting cinnamon with water. The active ingredient of cinnamon in terms of increasing insulin sensitivity, believed to be methylhydroxychalcone polymer (MHCP), is water soluble, but coumarin is not. Another way to reduce any worries about coumarin is to soak a cinnamon stick in tea. The tea will dissolve the MHCP but not the coumarin. And you even get the benefits of tea! Admittedly, not all studies have found that cinnamon is beneficial for diabetics, but in any case, diabetics are not the only ones who could benefit from daily consumption of cinnamon; anyone with high cholesterol can also give it a shot. But, of course, not in an apple pie!

VEGETABLES AND SALICYLIC ACID

"Organic food might reduce heart attacks." As you can imagine, that headline caught my attention, especially given that it appeared in *New Scientist*, a highly respected magazine. What prompted it, I wondered? Had researchers followed two groups of subjects, one feasting only on organic food, the other eating a conventional diet? And had they found a lower incidence of heart attacks in the former group? Not exactly. John Paterson and colleagues at the Dumfries and Galloway Royal Infirmary in Scotland analyzed the chemical composition of vegetable soups and discovered that the organic soups had six times as much salicylic acid as conventional soups. So what is the connection to heart disease? A clue is to be found in the effect of aspirin on the blood. It is fairly well established that a small daily dose of aspirin can offer protection against heart attacks by reducing the chance of blood-clot formation. Indeed, some physicians recommend that people over fifty consider taking a baby aspirin (eighty-one milligrams) every day. Chemically speaking, aspirin is acetylsalicylic acid, but in the body it breaks down to yield salicylic acid, which is the compound responsible for the physiological effects. It stands to reason, then, that the salicylic acid content of foods would be of interest to medical researchers.

A question immediately arises as to why vegetables contain salicylic acid in the first place. Obviously, plants did not evolve to produce substances that can protect humans against heart disease. But they did evolve to protect themselves against attacks by bacteria, fungi, or viruses. Salicylic acid serves as a plant hormone, activating genes that code for the production of proteins that battle the invaders. Organically grown vegetables, not protected by fungicides or pesticides, could be expected to have higher levels of salicylic acid, and according to Paterson's study, this seems to be the case. But before we start relying on organic vegetable soups to protect us against heart disease, we had better take a closer look at the numbers involved. The organic vegetable soups had an average of 120 nanograms of salicylic acid per gram of soup, while the conventional soups had 20 nanograms per gram. What does this mean? That a serving of organic soup, about 400 grams, contains roughly 0.06 milligrams of salicylic acid while the regular soup has 0.01 milligrams. Indeed, that is a sixfold difference. But now consider that a baby aspirin weighs in at 81 milligrams, which corresponds to more than a thousand times the amount of salicylic acid found in the organic soups. Clearly, this amount is irrelevant and choosing organic vegetable soups over others on this basis is pure folly.

Can the salicylic acid content of the diet ever be relevant? That's hard to say. Half a liter of red or white wine has about thirty milligrams of salicylic acid, which may be significant and at least partially responsible for the supposed protective effects of wine. As far as foods go, even the likes of tomatoes and apricots, which are regarded as high in salicylates, contain at most a couple of milligrams per serving. Obviously, one would have to eat a great deal of fruits and vegetables to get an appreciable amount of salicylic acid. Not a bad idea given that the benefits of such a diet would extend beyond protection against heart disease.

Colon cancer is common in the Western world but is rare in rural India. How come? Well, a clue may come from Westerners who habitually take aspirin for arthritis. Several studies have shown that aspirin may have a protective effect against colon cancer, although nobody suggests that it should be taken specifically for this purpose. In one widely publicized trial, colon cancer patients taking 325 milligrams of aspirin a day had a reduced risk of recurrence of the disease. By how much? For every ten patients treated with aspirin during the thirty-one months of the study, one recurrence was prevented. Not staggering, but still meaningful. It is aspirin's anti-inflammatory effect that is believed to offer protection against colon cancer. Is it possible that salicylic acid in the diet can have the same effect?

Let's turn to those rural Indians. A sampling of their blood serum shows a higher level of salicylic acid than normally found in the blood of Westerners. This is especially evident among Indians who are strict vegetarians. Indeed, the difference in blood salicylate levels can be as much as threefold. Buddhist monks, who totally abstain from meat, provide a particularly interesting example. Some have been found to have blood levels of salicylic acid comparable to those of people who take a daily dose of baby aspirin. Furthermore, crops grown in rural India probably have higher salicylic acid levels than equivalent crops grown in the West because they are raised without pesticides, herbicides, or fungicides. This means they are more likely to be attacked by pests, which in turn forces the plant to try to protect itself by synthesizing salicylic acid. Indians also use a large quantity and variety of spices in their cooking, and some of these are very rich in salicylic acid. Cumin, turmeric, chili powder, and paprika are great sources of this compound. And more importantly, the salicylic acid is readily absorbed. We know this because blood samples taken from volunteers after they have consumed a spicy meal rich in salicylates show an almost immediate rise in blood salicylic acid levels.

The lesson to take away from all this is that we yet have another reason to load up our plates with vegetables. But the salicylic acid picture is not rosy for everyone. As with almost all nutritional issues, there is a caveat. Some people have a sensitivity to salicylates. Asthma, skin rashes, and swelling of various body parts can occur in the very small percentage of the population that exhibits a sensitivity to these compounds. But as for me, I'll be sprinkling even more paprika into my vegetarian goulash. And I may even sneak in a little turmeric.

CARROTS AND CAROTENOIDS

The exploits of the Royal Air Force against the Luftwaffe during the Battle of Britain have become legendary. Why were British pilots so successful in downing German bombers? According to the Air Ministry, they gained their advantage by dining on carrots. This explanation sounded reasonable, even to German military intelligence. After all, scientists had long established that vitamin A deficiency could cause night blindness. Furthermore, it was known that beta-carotene, one of the orange-colored carotenoids found in carrots, could be converted by the body into vitamin A. If carrots could make the British see better in the dark, surely they would do the same for the German pilots. So the Luftwaffe ordered its pilots to eat carrots before their missions. But no matter how many carrots they wolfed down, they could not challenge British air superiority.

This was not surprising, since the Royal Air Force's success had nothing to do with carrots. The pilots' seemingly uncanny night vision was due not to vitamin A, but rather to a new invention called radar. The southern and eastern coasts of England had been lined with a chain of radar installations that could pinpoint the approaching German bombers for the RAF. The Air Ministry had actually cooked up the carrot story and fed it to German intelligence to send them searching for carrots instead of radar antennae.

Carrots may not have improved the pilots' eyesight, but recent research has revealed that beta-carotene does play a very significant role in maintaining health. This is probably due to its ability to act as an antioxidant and to neutralize free radicals. A Johns Hopkins University study of more than 25,000 people who had their blood sampled over a ten-year period supports this notion. Subjects with low beta-carotene levels had four times the rate of a certain form of lung cancer. The Western Electric Study in Chicago, which monitored the health status of 2,107 workers for nineteen years, also found that the incidence of lung cancer in smokers who had low carotene intakes was seven times greater than in smokers who ate a lot of high-carotene foods. At the Albert Einstein College of Medicine in New York, researchers found a threefold greater risk of cervical cancer in women with low carotene intake.

There are also interesting links between beta-carotene intake and heart disease. Some 22,000 doctors enrolled in the Physicians' Health Study (begun in the fall of 1982 to test the benefits and risks of aspirin and beta-carotene in the primary prevention of cardiovascular disease and cancer) were asked to take either a fifty-milligram beta-carotene tablet or a placebo every second day. While no significant differences in cancer rates were noted, supplements did cut the heart attack risk in half among the subjects who had signs of heart disease when they entered the study.

The largest long-term study of women in the world is the Nurses' Health Study, coordinated by Harvard Medical School. Over the period of the study, women consuming a diet containing more than fifteen to twenty milligrams of beta-carotene daily had a 40 percent reduced risk of stroke and a 22 percent reduced risk of heart attack compared with women taking less than six milligrams of beta-carotene. In 1,000 women who had angina, the highest carotene consumers had an 80 percent reduced risk of heart attack.

These studies were highly publicized in the lay press, and it came as no surprise that many people started downing beta-carotene supplements. But the bandwagon screeched to a halt in 1994 with the revelation that in a Finnish study smokers who took beta-carotene supplements actually developed *more* cases of lung cancer. Critics tried to pass off these findings as anomalous but were quieted when an American study of smokers also showed an almost 30 percent increase in lung cancer among subjects taking daily thirty-milligram beta-carotene supplements. What's going on here?

Researchers at Tufts University tried to "ferret out" the problem. They fed high doses of beta-carotene to the weasel-like animals, which metabolize the compound the same way humans do. Some of the ferrets also inhaled an amount of smoke equivalent to thirty cigarettes a day for six months. The incidence of lung tumors increased, especially among the smoking ferrets. But an analysis of the animals' blood suggested a solution to the paradox. At high levels, beta-carotene actually acts as an oxidant instead of as an antioxidant!

Beta-carotene's antioxidant effect can be ascribed to the fact that it can neutralize free radicals by donating an electron. In the process, however, beta-carotene itself becomes a free radical that can damage tissue unless it is appeased by some other molecule from which it can snatch an electron. This is where vitamins E and C enter the picture. These compounds are very adept at scavenging the carotene radical without generating other dangerous free radicals. Since smokers are known to have low blood levels of vitamin C, they can be expected to be at greater risk from beta-carotene supplements.

More evidence for the unusual behavior of beta-carotene comes from, of all things, chicken feed. Fat is commonly added to improve the efficiency of agricultural feeds. Unsaturated fat is better because it improves the nutritional profile of the final product, but unfortunately, unsaturated fats in the meat oxidize more easily than saturated fats, degrading taste and texture. Producers have experimented with fortifying the feed with vitamin E and beta-carotene to reduce oxidation. They discovered that when beta-carotene was added it behaved as an oxidant unless vitamin E was also added. With sufficient vitamin E, however, beta-carotene exerted its expected antioxidant effect.

So what are we to do with this information? For the moment, it is probably advisable to lay off beta-carotene supplements but not to reduce our consumption of foods rich in beta-carotene. That's because beta-carotene may require the presence of other food components to exercise its benefits. There is no recommended daily intake of beta-carotene, but a review of the literature reveals that we should strive for about twenty to twenty-five milligrams daily. To put this amount into perspective, note that a sweet potato has about fifteen milligrams, a carrot twelve milligrams, half a cantaloupe five milligrams, a half cup of spinach four milligrams, and a spear of broccoli two milligrams.

So we've seen the benefits of beta-carotene. And talking about seeing. . . . There is one final story, and it has nothing to do with night blindness. It has to do with cataracts, the leading cause of blindness around the world. As we age, free-radical reactions cause the protein in the lens of the eye to clump and to form the opaque deposits we call cataracts. Light entering the lens scatters before it can get through to the retina. Several recent studies have shown that a high intake of antioxidant nutrients, particularly carotenoids, is associated with a decreased risk of cataract formation.

Carrots may not have defeated the Germans but they may help us win the war against cancer and heart disease. And they may even help us see the future more clearly.

VITAMINS FROM A TO K

The basic definition of vitamins is pretty straightforward. They are substances that must be included in the diet in order to maintain health and prevent certain deficiency diseases. What sort of deficiency diseases are we talking about? Scurvy was the first one recognized, described as early as 1550 BC by the Egyptians in the Ebers Papyrus. In the sixteenth and seventeenth centuries, when long ocean voyages became common, thousands of sailors died from the disease. The first clue toward solving the problem came when French explorer Jacques Cartier's ships became icebound in Quebec in 1536. Only three of his 100 men escaped the ravages of the scurvy. It was then that the native Stadacona people came to his rescue and advised the men to make a tea by boiling the leaves of a tree, probably the white cedar. The men recovered rapidly after only a couple of doses, but the remedy seems to have been lost. There were other instances of effective scurvy treatment. In the seventeenth century, some ships of the East India Company carried supplies of lemon juice to ward off the disease. Still, these were isolated cases, and thousands of sailors continued to perish from scurvy.

Scottish physician James Lind had heard accounts of treating scurvy with various foods or beverages and decided to get to the bottom of the matter. Aboard HMS *Salisbury*, he selected six pairs of men with scurvy. To each pair he gave one of the following daily doses: cider; dilute sulphuric acid; vinegar; sea water; a mash of garlic, mustard seed, and radish root; or two oranges and a lemon.

There was also a control group of men with scurvy who got the regular ship's rations. Within days, the two men who were lucky enough to have been put on the citrus diet began to recover. So although Lind was not the first to discover a treatment for scurvy, he certainly was the first to document a "clinical trial" showing the effectiveness of the citrus remedy, which he did in his "Treatise on Scurvy" in 1753. Still, it wasn't until 1795 that the Royal Navy began to provide a daily supply of lime or lemon juice to all its men, thereby giving rise to the expression "Limey" to describe natives of Britain. Around the same time, Captain James Cook discovered that fresh fruits and sauerkraut were also scurvy preventatives. Finally, in the 1930s, Albert Szent Györgyi isolated the scurvy-protective factor and named it vitamin C. Why? Because the idea of naming vitamins by letters had already been introduced some twenty years earlier and A and B had been taken.

Letter designations for vitamins go back to the early part of the twentieth century. When the mechanized rice mill was introduced in Asia, a new disease that came to be called "beriberi" cropped up. Beriberi means "I can't, I can't" in Sinhalese, a native language of Sri Lanka, and describes a condition of progressive muscular degeneration, heart irregularities, and emaciation. Kanehiro Takaki, a Japanese medical officer, studied the high incidence of the disease among sailors in the Japanese navy from 1878 to 1883. He discovered that on a ship where the diet was mostly polished rice, among 276 men, 169 cases of beriberi developed and 25 men died during a nine-month period. On another ship, there were no deaths and only fourteen cases of the disease. The difference was that the men on the second ship were given more meat, milk, and vegetables. Takaki thought this had something to do with the protein content of the diet, but he was wrong.

About fifteen years later, a Dutch physician in the East Indies, Christiaan Eijkman, noted that chickens fed mostly polished rice also contracted beriberi but recovered when fed rice polishings. He thought that the starch in the polished rice was toxic to the nerves, but he too was wrong. Finally, Casimir Funk, a Polish chemist, got it right. He showed that an extract of rice hulls prevented beriberi. Believing that this substance fell into the chemical category of amines, and since it was vital to life, he called it "vitamine." When the substance turned out not to be an amine, the *e* was dropped.

A short time later, E. V. McCollum and Marguerite Davis at the University of Wisconsin discovered that rats given lard as their only source of fat failed to grow and developed eye problems. When butterfat or an ether extract of egg yolk was added to the diet, growth resumed and the eye condition was corrected. McCollum suggested that whatever was present in the ether extract be called fat-soluble factor A and that the water extract Funk had used to prevent beriberi be called water-soluble factor B. When the water-soluble extract was found to be a mixture of compounds, its components were given designations with numerical subscripts. The specific anti-beriberi factor was eventually called vitamin B_1, or thiamine. These vitamins had a common function. They formed part of the various enzyme systems needed to metabolize proteins, carbohydrates, and fats. Some of the compounds in Funk's water extract eventually turned out to offer no protection against any specific disease, and their names had to be removed from the list of vitamins. As other water-soluble substances required by the body were discovered, they were added to the B-vitamin list.

Other vitamins were subsequently identified and given the designations D and E in order of their discovery. Vitamin K was so called because its discoverer, the Danish biochemist Henrik Dam, proposed the term *koagulations vitamin* due to its promotion of blood coagulation. Are there still unrecognized vitamins? Not likely. Patients have now been kept alive for many years through total parenteral nutrition (TPN), which involves using an intravenous formula that incorporates the known vitamins. While it is most unlikely that new vitamins will be discovered, it is certainly possible that new uses for vitamins will come to the fore. We now recognize that vitamins can do more than prevent the classic nutritional deficiency diseases. They may play a role in warding off heart disease, cancer, and perhaps even Alzheimer's disease.

SPINACH AND THE B VITAMINS

The most famous landmark in Crystal City, Texas, is a statue of Popeye the Sailor Man. He's squeezing his trademark can of spinach, ready to save Olive Oyl from the clutches of Bluto. Crystal City, you should know, is the spinach capital of the world. Its citizens erected the statue in 1937 to honor the man who single-handedly boosted spinach consumption and helped save an industry. And Popeye may have done more than give Crystal City an economic boost. He may also have contributed to improved health for its citizens. That's because spinach is an outstanding source of folic acid, a B vitamin that is increasingly being linked with a plethora of health benefits.

Our story starts in the hallowed halls of Harvard University, far from the spinach fields of Crystal City. It was here in 1969 that Dr. Kilmer McCully got involved in the unusual case of a boy who died at the age of eight from a stroke. He had suffered from a rare condition that caused the buildup of a substance known as homocysteine in his blood. This is a normal metabolite of methionine, a common amino acid found in virtually all dietary proteins. A healthy person's body quickly processes homocysteine, but it accumulates in those suffering from homocystinuria, like McCully's young patient. An autopsy was performed and it clearly revealed the cause of death. The boy's arteries were like those of an old man! Could the damage have been caused by excess homocysteine, McCully wondered? To investigate this further, he needed to examine other children who were afflicted with the same condition.

It didn't take long for Dr. McCully to reach a conclusion. Children with high homocysteine showed artery damage typical of that seen in older men. To prove his point, he injected homocysteine into rabbits and triggered artery damage. This was enough evidence to suggest a revolutionary idea: homocysteine was a risk factor for heart disease. McCully proposed that high levels caused damage quickly while levels that were only slightly elevated took a longer time to wreak havoc. Excited about his findings, he submitted a paper for publication to the *American Journal of Pathology*. But instead of getting fame, he got fired.

Harvard denied McCully tenure, supposedly because of his unorthodox theory about heart disease. The medical establishment had declared that cholesterol was the main culprit, and there seemed to be no room for homocysteine. Dr. McCully would, however, be eventually vindicated. At least for a while. Somewhat fittingly, one of the first studies to show the possible validity of the homocysteine theory was carried out at the Harvard University School of Public Health. In 1992, researchers reported on an analysis of disease patterns in over 14,000 male physicians. Those whose blood levels of homocysteine were in the top 5 percent had a heart attack risk that was three times greater than that seen in subjects with the lowest levels. Numerous other studies have shown a similar relationship. High homocysteine seems to be a clear, independent risk factor for heart disease.

Knowing about a risk factor is not much good unless something can be done about it. And in the case of homocysteine it can. Let's take a moment to explore the relevant biochemistry. Homocysteine forms through the action of certain enzymes on methionine. Once it has formed, one of two things will happen. It is either reconverted to methionine or metabolized to glutathione, a powerful antioxidant. Both of these pathways require the presence of B vitamins. Folic acid and vitamin B_{12} are needed to change homocysteine back to methionine, and vitamin B_6 is required for the glutathione route. You are probably starting to get the picture. Inadequate levels of these B vitamins lead to increased circulating homocysteine, creating a risk factor for heart disease. But to prove that elevated homocysteine causes heart disease, intervention studies were needed.

The Heart Outcomes Project Evaluation (HOPE) was designed to determine the effect on heart attack or stroke of lowering homocysteine. Over 5,000 patients at risk due to existing vascular disease or diabetes were given B vitamins or a placebo. After five years, the subjects who took 2.5 milligrams of folic acid, 50 milligrams of vitamin B_6, and 1 milligram of vitamin B_{12} daily were no better off than those who took a placebo. These results occurred in spite of a 25 percent reduction in blood homocysteine. A Norwegian study that involved giving B vitamins to men and women after a heart attack came to the same conclusion. Again, homocysteine was reduced, but the risk of a second heart attack or sudden death was not. Homocysteine, it seems, may signal the approach of heart disease, but it doesn't cause it. So McCully's homocysteine theory about heart disease is not on as firm a footing as it once seemed. But there is more to the folic acid saga.

A recent study of 25,000 women showed that those who consumed the most folic acid were one-third less likely to develop precancerous polyps in their colons. And if that isn't motivation enough to seek out foods high in folic acid, then consider that it may even lower the risk of Alzheimer's disease. Researchers at the University of Kentucky explored the Alzheimer's connection because they were aware of the extensive evidence showing that folic-acid supplements during pregnancy can help prevent neurological birth defects such as spina bifida. Could folic acid affect the nervous system later in life, they wondered? A group of nuns in Minnesota who had willed their bodies to scientific research provided the answer. Nuns who had an adequate intake of folic acid throughout their lifetimes were less likely to succumb to Alzheimer's. This finding was corroborated by researchers at Tufts University who fed spinach to rats and found that it not only prevented but reversed memory loss. But once again, when it comes to human intervention trials, the results are ambiguous. When close to 300 healthy elderly people with high homocysteine levels were given a daily supplement of 1,000 micrograms of folic acid along with 500 micrograms of vitamin B_{12} and 10 milligrams of vitamin B_6, researchers were unable to detect any difference in cognitive performance when compared with a control group. On the other hand, Jane Durga of Wageningen University in the Netherlands found that giving elderly adults with elevated homocysteine levels 800 micrograms of folic acid a day improved cognitive function significantly.

The B vitamins have a very good safety profile and the doses needed to keep homocysteine in check are not extreme. About 400 micrograms of folic acid, 3 micrograms of B_{12}, and 3 milligrams of B_6 daily should do the job. While it is certainly possible to get these from the diet, the fact is that many people just don't. Indeed, the average intake of folic acid in North America is about 200 micrograms, probably far from adequate. That's why a supplement may be appropriate. But you don't want to overdo it. As with so many substances, more is not necessarily better. The previously described Nurses' Health Study, which suggested that women with the highest folic acid intake were least likely to get colon cancer, stimulated a study at Dartmouth College of more than one thousand men and women who had previously had polyps removed from their colon. They were given 1,000 micrograms of folic acid a day, in the hope of preventing cancer. Not only did that not happen, but there was actually a slight increase in cancerous polyps. But this study involved people who already had polyps and therefore says nothing about folic acid preventing polyps in the first place, as was evidenced in the Nurses' Health Study. Another disturbing finding of this study was an increased incidence of prostate cancer. But we have to keep in mind that the supplemental dose of 1,000 micrograms is considerably larger than the 400 micrograms found in most supplements. What are we to conclude from all this? That the effects of folic acid supplementation on babies may be different from that on adults, and that in adults folic acid may prevent new cancers but make existing cancers worse. At this point, there is no reason to take any supplement that has more than 400 micrograms of folic acid.

But eating spinach is safe enough. It is an outstanding source of folic acid, particularly if it is not cooked. So go for that spinach salad! May I suggest dressing it with orange juice? Just one cup of OJ contains 100 micrograms of folic acid. You can also throw in some green beans or cooked brown beans or asparagus, also great sources of folate. And if you can't remember all this, well, you probably need to eat more folic acid.

OILS, NUTS, WHOLE GRAINS, AND VITAMIN E

Over half a century ago, Dr. Evan Shute and Dr. Wilfrid Shute of London, Ontario, thought they had taken a giant step toward solving the problem of heart disease. Just 200 IU of vitamin E a day, they claimed, was beneficial in reversing heart disease and in treating angina. The Shutes followed thousands of patients, gathered data, and submitted papers to medical journals. "Anecdotal," "no controls," "poorly designed trials," said the editors as they rejected the manuscripts. But word about the supposed successes with vitamin E spread and so did the popularity of supplements, despite the skepticism of the medical community. Since the time of the Shutes, numerous studies have been carried out on vitamin E, and you would think that by now we would have a pretty good grasp on whether or not to recommend supplements. Alas, such is not the case. After the plethora of research, only two certainties have come to light. One, vitamin E may behave as an antioxidant, and two, it is no panacea.

Any discussion of this celebrated vitamin really should begin with a description of its chemistry. Just what is vitamin E? Almost immediately we run into a problem, because the answer to this question is not so simple. Unlike vitamin C, for example, vitamin E is not one single compound, and again unlike vitamin C, the synthetic version is not identical to the natural version. But let's start at the beginning. Back in the 1920s, researchers noted that male rats lacking fat in their diet became sterile and that females failed to carry their young to full term. Eventually, the problem was traced to a fat-soluble substance for which the term *tocopherol* was coined, deriving from the Greek *tokos* for "birth" and *pheroi*, "to carry." Since the substance could not be made in a rat's body and had to be supplied by the diet, it fit the definition of a vitamin and tocopherol became vitamin E.

It didn't take long for the first issue about vitamin E to arise. Chemical analysis revealed that vitamin E wasn't a single compound; there were actually eight closely related substances that had "vitamin E activity." The differences in molecular structure were subtle, but nevertheless did result in different physiological effects. D-alpha tocopherol turned out to have the greatest biological activity, as determined by its effectiveness in preventing reproductive problems in rats. Chemists soon learned how to isolate this specific form of vitamin E from natural products such as wheat germ and soybeans, and "natural" vitamin E supplements hit the marketplace. Not only did the clever chemists learn to isolate d-alpha tocopherol, they also figured out how to synthesize it in the lab. But there was a slight difference here. When it was made in the lab, the compound inevitably formed together with its nonidentical mirror image form, l-alpha tocopherol, which did not exist in nature. The *l-isomer*, as it was called, had far less biological activity than the *d* version.

Now, though, there was a problem. Since the eight naturally occurring components of vitamin E and the synthetic *l* version all had different biological activities, there was a need for some standardized unit of measure for vitamin E activity. Weight would be misleading because one milligram of synthetic vitamin E, which was composed of the active *d* and the less active *l* forms, would not have the same effect as one milligram of pure *d*. Therefore the term International Unit (IU) was defined to represent the biological activity of one milligram of synthetic vitamin E. By this scale, d-alpha tocopherol has an activity of 1.49 IU. Therefore all tablets labelled as having 200 IU of vitamin E have exactly the same ability to prevent reproductive problems in rats, although they may not have exactly the same composition.

Most forms of "natural" vitamin E contain d-alpha tocopherol extracted from soybeans, although there are some versions available that contain all eight vitamin E components. The "synthetic" version consists of equal amounts of d-alpha tocopherol and l-alpha tocopherol. Of course, today, interest isn't focused on vitamin E's effect on reproduction; the various health claims made on its behalf are what intrigue people. And there is no shortage of these. The Shutes' original anti-heart-disease claims have been joined by a host of others. Vitamin E supposedly enhances longevity, reduces the risk of Parkinson's and Alzheimer's disease, has anticancer properties, protects the prostate and, depending on whom you listen to, is beneficial for almost any other human ailment. The vitamin's only shortcoming seems to be a lack of hard corroborative evidence.

By the late twentieth century researchers had shown that, at least in the lab, vitamin E had the capability of neutralizing free radicals. That was an encouraging finding, given that free radicals are implicated in a number of diseases, and it seemed to mesh with the epidemiological studies that demonstrated a reduced incidence of heart attack and stroke in people taking vitamin E supplements. It seemed clear sailing for vitamin E, except for one little annoyance. At high doses the vitamin had an anticoagulant effect, but this was not an issue at the 200 IU to 400 IU level most people were taking. Still, there was the lingering suspicion that people taking vitamin E were healthier not because of the vitamin but because they were more likely to follow a healthy lifestyle. This question would be resolved, scientists thought, with proper intervention studies. Give some subjects vitamin E, others a placebo, monitor them for years, and see what happens.

A number of such intervention studies have now been carried out and the results published. The hoped-for benefits of vitamin E have not materialized. People taking vitamin E fared no better in terms of heart disease than those taking a placebo. When Dr. Edgar Miller of Johns Hopkins University pooled together data from the best vitamin E studies in a meta-analysis, he found a startling result. Not only did the vitamin not protect against disease, it appeared to increase mortality! Consumers were shocked. Supplement manufacturers went into a spin frenzy and suggested that most of the studies involved people who already had cancer, Alzheimer's, or heart disease, and therefore the results would not be expected to apply to a healthy population. Not really a valid criticism. Actually, the greatest effect would be expected in people who already have some disease. Aspirin, for example, is of great use in preventing heart attacks in people with existing heart disease, but the jury is still out on the effectiveness of healthy people taking aspirin. So if vitamin E offers no help to those who suffer from an ailment, it is unlikely to benefit the healthy.

Criticism was also aimed at the fact that Miller had not separated studies that used natural vitamin E from those that used the synthetic version. Benefits are more likely from the natural vitamin, some say. Actually, the difference between these is very subtle and is taken care of by standardizing doses in International Units. But even strong supplement supporters have a hard time getting around the fact that no benefit was seen with vitamin E use in 136,000 people, and that there was a dose-response relationship in terms of mortality. In general, when an effect, either positive or negative, increases with dose, it usually means that it is real rather than a statistical artifact. The vitamin E meta-analysis implied that risk of premature death begins to rise at a daily dose of around 150 IU of vitamin E, and that at a dose of 400 IU per day the risk of dying from any cause becomes about 10 percent higher than for people not taking the vitamin. Perhaps vitamin takers don't take as much care with their diet and exercise habits because they feel they are protected, but this is not a likely explanation, given the large number of subjects involved in the studies. The fact is that as more and more high-quality studies about supplements come to light, we begin to see an emerging pattern. While antioxidants undoubtedly play an important role in health, their relative amounts are of the essence. More is not necessarily better. Food seems to contain the best balance of these nutrients, and when we flood the body with antioxidants from an outside source, the antioxidant balance is upset to the extent that adverse reactions may occur.

The book on vitamin E is not closed, however. There are indications that it may play a role in preventing Parkinson's disease, that it may work against cold sores, and that an inadequate intake during pregnancy may raise a child's risk of asthma. And if you want to train mice to be circus performers, you better make sure they have enough vitamin E in their diet. That's if you go by the work of researchers at the University of Cádiz in Spain and at the University of Buenos Aires in Argentina. Ana Novarro and Alberto Boveris studied the acrobatic prowess of mice on the high wire—high at least for mice. They stretched a tightrope half a meter above the ground and made the animals scamper across it, keeping a careful eye on just how well the creatures were able to maintain their balance. The rodents' performance was evaluated regularly over a period of some sixty weeks. The researchers were not auditioning for novelty acts; they were interested in studying the effect of vitamin E intake on the mice's coordination.

The 300 animals in the study were fed normal laboratory chow, but half of them received a daily vitamin E supplement. And guess what? All the mice lost some of their balancing ability with age, but still, at the ripe old age of seventy-eight weeks, the animals fed vitamin E performed some 45 percent better. The mice were given the human equivalent of about 2,000 IU of vitamin E, which is far above the upper limit recommended by most nutritional authorities. Because of the high dose used, we might be tempted to dismiss this study, especially since most seniors do not regard tightrope walking as a necessary facet of their lives. But wait! The researchers found that not only did the vitamin E supplements help the mice with their agility on the tightrope, but that the supplemented animals also lived about 40 percent longer than expected. Excited by this result, the scientists autopsied the mice to see if they could find any molecular evidence for the apparent anti-aging effect. And they did! It is well known that aging is accompanied by cellular damage attributed to the action of free radicals. In the case of the vitamin E mice, fewer compounds that are the product of free-radical damage were found, particularly in the animals' brains.

Can vitamin E prevent cognitive decline among humans? Well, we don't have any human tightrope-walking studies, but we do have some interesting results generated by the work of Martha Clare Morris at Rush University Medical Center in Chicago. Dr. Morris had about 3,700 seniors with an average age of seventy-four fill out extensive food-frequency questionnaires, and she evaluated their mental status by administering four different tests. The exercise was repeated three years later and again after another three years. Her conclusion? Compared with people who consumed less than one serving of vegetables a day, people who ate at least three servings saw their rate of cognitive change slow by roughly 40 percent. Leafy green vegetables had the strongest association with protection, but surprisingly fruits showed no effect. When the researchers attempted to correlate various components known to be present in vegetables to better mental performance, vitamin E stood out. And that observation could explain the lack of a protective effect from fruits. Vegetables are often consumed with some sort of fat, as in salad dressings, and fat is known to increase the absorption of vitamin E. If you needed any more evidence to increase your vegetable intake, this study provides it.

Since we have no clear indication of benefit from high doses of vitamin E, and since there are suggestions of possible harm, the prudent advice is to avoid high doses. It is unlikely that doses up to 400 IU are harmful, but it is better to get our vitamin E from foods such as leafy green vegetables, nuts, and whole grains. Although there are those who claim various benefits from taking vitamin E supplements, their views are not supported by high-quality, placebo-controlled, randomized trials.

COD LIVER OIL AND VITAMIN D

As the Industrial Revolution took hold in Britain, physicians began to notice an unusual phenomenon: many children were becoming bowlegged. Their bones, it seemed, were just too weak to support their body weight. Nobody at the time realized that rickets, as the condition came to be called, was caused by a lack of sunshine. England's skies were filled with dense, black smoke spewing out from the mushrooming factories, often obliterating the sun. This greatly reduced exposure to the ultraviolet light needed to produce vitamin D, in the body. This vitamin plays a critical role in the absorption of calcium, necessary for bone formation. The connection between sunshine, vitamin D, and rickets, however, was not made until the early years of the twentieth century. It was then that Dr. Alfred Hess and Dr. Mildred Weinstock of Columbia University laid the foundation for our understanding of the role of vitamin D in bone formation, with an ingenious experiment.

The Columbia researchers deprived rats of sunshine until they developed rickets. Then they excised a piece of skin from the animals, placed it in brilliant sunshine and then added it to the rats' food. Lo and behold, the rodents quickly recovered from their affliction. The sunshine had triggered the production of some sort of antirickets factor in the exposed skin. Around the same time in England, Dr. Edward Mellanby added another piece to the puzzle. Mellanby, a lecturer at King's College for Women in London, thought that rickets might be due to some dietary deficiency. The British diet at the time lacked variety, and many of the poor subsisted on little other than porridge.

Mellanby decided to feed a bunch of dogs nothing but oats, and to his great satisfaction they got rickets. This convinced him that some substance necessary to prevent rickets was missing from oats.

Since not everyone who ate porridge was affected by rickets, the mysterious substance must be present in other foods, Mellanby concluded. Maybe the key to a bone-healthy diet was consumption of animal products. For some bizarre reason, he decided to try adding cod liver oil to the dogs' diet. To his amazement, the animals were cured of rickets. Soon cod liver oil was being forced down the throats of wriggling children all over Britain, and rickets essentially became a relic of an earlier age. Mellanby had hit upon a cure, and amazingly he had done so with a faulty conclusion. Oats had nothing to do with the onset of rickets. Inadvertently, the dogs in Mellanby's experiment had been kept in the dark, and it was the lack of light that caused the disease. But Mellanby was certainly correct in assuming that cod liver oil contained a substance that could cure rickets. That substance of course turned out to be vitamin D.

Our bones are essentially composed of calcium phosphate, the components of which originate in the diet. But the absorption of calcium from the digestive tract requires the presence of a transport protein. This is where vitamin D enters the scene; the transport protein cannot be made without it. And to make things even more confusing, a specific form of vitamin D, namely 1,25-dihydroxyvitamin D_3, is needed. This is not the form that is found in foods or is formed upon exposure to sunlight. Vitamin D_3, the form synthesized in the skin on sun exposure, is converted to 25-hydroxyvitamin D_3 in the liver, and then is modified to the active form in the kidney.

As soon as the link between vitamin D and rickets came to light, the idea of fortifying foods with the vitamin emerged. But this required large-scale production of the vitamin, a challenge that was met in an ingenious fashion by exposing cow, pig, or sheepskin to sunlight and extracting the vitamin D_3 that formed with a solvent. Milk, since it contained calcium, was chosen as the ideal vehicle to increase vitamin D intake. By the 1940s, widespread milk fortification was in place and the incidence of rickets was practically eliminated.

Today we face a different problem. Fear of the sun has resulted in low blood levels of vitamin D in adults, particularly in seniors. The levels are not low enough to cause rickets, but they can cause a softening of the bones (osteomalacia) or, in extreme cases, bone-brittling osteoporosis. During the winter months in the Northern Hemisphere, the effective wavelengths of sunlight do not penetrate the atmosphere and vitamin D supplements have to be considered. There is a question, however, about just how much vitamin D we need. The usual recommendation has been 200 IU daily for people under fifty; 400 IU for those between fifty and seventy; and 600 IU for people over seventy. Many researchers, however, think that these recommendations should be increased, given that studies have shown that an intake of 1,000 IU a day is needed to reduce the risk of fractures.

But vitamin D may do more than reduce the risk of fractures. In the 1940s, Dr. Frank Apperley, in a landmark paper in the journal *Cancer Research*, reported that human death rates from various cancers increased proportionally according to people's distance from the equator. Apperley wondered whether the effect of sunshine could account for this observation. By 1980, researchers had confirmed that colon, breast, and prostate cancers were more common at latitudes with less annual sun exposure. Of course, it is important not to jump to conclusions here, because dietary patterns and activity levels could account for the difference. Still, the vitamin D connection is intriguing because receptors for this hormone have been found in cells in many different organs, including the breast, the prostate, and even the brain. So vitamin D seems to do a lot more than affect bone formation.

Led by Dr. Cedric Garland, researchers at the University of California confirmed this finding. When blood samples were taken from 701 breast cancer patients and compared with samples from a similar group of healthy women, the researchers discovered that a high blood level of vitamin D was associated with a reduced risk of breast cancer. But to achieve the levels that seem to offer protection from the disease, people would need to take a daily dose of 1,000 IU, an amount very few people get. Such doses can be achieved only through dietary supplementation, not from sun exposure.

This doesn't mean the sun doesn't play an important role. A Canadian study carried out at Mount Sinai Hospital in Toronto compared the history of 1,000 breast cancer patients with the history of cancer-free matched controls. Questionnaires filled out by the women revealed that those who spent more time in the sun—especially as teenagers—as well as those who had a high intake of vitamin D from their diet—about ten glasses of fortified milk a week when they were young—were significantly less likely to develop breast cancer later in life. The researchers observed an approximate reduction in risk of 30 percent. It seems the critical time for a high vitamin D intake is while the breast tissue is being formed. After age forty-five, no link to protection from cancer was found.

Perhaps the most convincing evidence for the potential benefits of supplementing the diet with vitamin D comes from a meta-analysis of vitamin D trials published in the *Archives of Internal Medicine* in September of 2007. This "study of studies" received widespread press coverage, often featuring headlines such as "Vitamin D Reduces Risk of Death by 7 Percent." This, of course, is a headline writer's hyperbole. Vitamin D will not help anyone live forever. What the researchers actually showed was that subjects taking vitamin D supplements were less likely to die from any cause than subjects taking a placebo. Overall, the researchers looked at eighteen studies that investigated the effect of vitamin D on bone fractures, cancer, and heart disease. While none of these studies were designed to study mortality, all of them recorded the deaths of subjects involved. Pooling of the results led to the conclusion that people who took an average of approximately 500 IU of vitamin D a day in the form of supplements were 7 percent less likely to die during the study.

Evidence is also accumulating that vitamin D may forestall diabetes in people at risk, that it enhances immune function, and that it slows the progression of osteoarthritis. But all of these effects probably require dosages greater than the current recommendations. The best guess is that 1,000 IU a day is a good target. Even at 2,000 to 3,000 IU a day there appears to be no risk, except perhaps in people prone to kidney stones. Vitamin D supplements may turn out to be one of those rare cases in which at least some of the advertising hype is justified.

MILK AND CALCIUM

You would think that if there is one food item incapable of stirring up a health debate, it would be milk. After all, it is the only substance we consume that evolved with the sole purpose of serving as a food. But to say that milk stirs a debate is to put it mildly. Arguments about its benefits or harm provoke downright vicious battles that transcend nutritional controversies. In one corner, we have organizations such as the Physicians' Committee for Responsible Medicine (PCRM), the Anti-Dairy Coalition, and People for the Ethical Treatment of Animals (PETA) that maintain milk is a deadly poison and that "cow's milk is for calves." In the other corner, we have the Dairy Association and various independent researchers who claim that drinking milk contributes significantly to good health. Both sides vie to convince the public of the "truth" with extensive and expensive advertising campaigns, backed by references to the scientific literature.

Of course, there is more than science involved here. The Dairy Association looks out for milk producers and does what it can to promote the sale of milk and milk products. The antimilk groups use the issue to further their animal rights and vegetarian agendas. It would seem that the two combatants have nothing in common, but they do. Both jump on any study that backs their cause and immediately dismiss any that does not. And with the plethora of studies being cranked out these days, "evidence" can be found for almost any view. Responsible science, though, involves casting away agendas, taking off blinders, and scrutinizing the totality of evidence before coming to a conclusion.

Milk stands accused of contributing to heart disease, stroke, breast cancer, prostate cancer, ovarian cancer, diabetes, allergies, stomach cramps, diarrhea, autism, mucus production and, get this, bone fractures! But milk is also linked with reducing heart disease, breast cancer, colorectal cancer and, of course, bone fractures. It all depends on whom you listen to. The antimilk arguments often begin with the observation that no other species except humans drink milk after weaning. That's hardly a compelling argument. No other species designs airplanes, develops antibiotics, or bakes bread either.

Heart disease is indeed more prevalent in countries where dairy consumption is high. But these countries have a diet high in total saturated fat. Yes, milk contains saturated fat, but what matters is the sum of all the fat consumed. The saturated fat in milk can be avoided by consuming low-fat dairy products. It is interesting to note that in a recent study, Professor Peter Elwood of the University of Cardiff in Wales tracked 400,000 adults worldwide for twenty-eight years and found that those who drank the most milk had a lower risk of heart disease and stroke than those who drank little or no milk. And no, he did not have funding from the dairy industry. Elwood's is not the only study to find such a result. At the University of Bristol, researchers had 764 men weigh and record every item of food and drink they consumed for a week. Then the men were followed for twenty years. High-quantity milk drinkers had a lower risk of heart disease and stroke than those who drank the least. It has been suggested that calcium's ability to reduce blood pressure may be involved. Calcium is known to increase the rate at which the body produces nitric oxide, a chemical instrumental in relaxing the walls of blood vessels, thereby lowering blood pressure.

Are dairy products a risk factor for breast or prostate cancer? There is a worldwide increase in such hormone-related cancers, and cow's milk is a source of estrogens. Modern dairy cows are usually pregnant and are milked during their pregnancy when estrogen concentrations are high. Dairy foods also contain an insulin-like growth factor (IGF-1), which can cause irregular cell multiplication. Furthermore, milk contains traces of dioxins that originate from pollutants that may have settled on fields where the animals graze. And dairy products are rich in calcium, which of course builds bones, but also depletes blood levels of a form of vitamin D that has been linked with protection against cancer.

That's the theory, but what does the epidemiological evidence say? A number of studies show a correlation between the incidence of breast and prostate cancer and dairy consumption. But the correlations tend to vanish when an adjustment is made for nonmilk fat intake. Animal-derived fats in general may have an adverse effect on hormone-dependent cancers, but milk is not specifically implicated. In the case of prostate cancer, studies have shown a link to calcium intake, but not to overall dairy intake. A couple of glasses of milk a day is not a problem, and calcium in such amounts has been shown to offer protection against colon cancer.

There is even some evidence that milk can actually reduce the incidence of breast cancer. Consuming more than nine glasses of milk per week compared with fewer than five glasses, at ages twenty to twenty-nine, has been linked with protection against the disease. A Finnish study tracked close to 5,000 women for twenty-five years and found that those who consumed the most whole milk had the lowest incidence of breast cancer. Compounds called conjugated linoleic acids (CLAs) may be protective and have actually been shown to suppress breast tumors in animals. As far as ovarian cancer goes, milk drinking does increase the risk somewhat, but this is outweighed by the reduced risk of colorectal cancers, which are far more common.

A few years ago, a study that suggested a link between milk and type 1 diabetes caused a lot of commotion, but it has never been corroborated. Nor is there any evidence that milk causes mucus, although people who have an allergy to milk can become congested. But without a doubt, people who suffer from lactose intolerance can develop gastrointestinal symptoms. This condition results from the inability to digest lactose, a sugar found in milk. About 70 percent of the world's population lacks the ability to produce the enzyme beta-galactosidase (better known as lactase), which is essential for proper digestion of lactose. Lactose intolerance is most prevalent among people of Asian, African, and, to a lesser extent, Mediterranean origins. Many parts of Asia and Africa were once afflicted by sleeping sickness (African trypanosomiasis), a disease transmitted by the tsetse fly and responsible for the destruction of cattle populations. Geneticists believe that the resulting unavailability of milk in these areas led to lactose intolerance. In terms of evolution, such a response would be appropriate, since disruption of the synthesis of a nonrequired enzyme would be advantageous to the human body. Although lactose intolerance is rare among infants, the ability to produce the enzyme drastically decreases in predisposed individuals during the years after weaning. Many lactose-intolerant people can still consume small amounts of milk without the serious ill effects normally associated with the condition, namely diarrhea and abdominal cramps.

The diarrhea is likely the result of increased flow of water into the intestine (through osmosis) in response to lactose buildup. At the same time, fermentation of small amounts of lactose by bacteria commonly found in the digestive tract results in the production of gases that can lead to cramps. One of the gases produced, hydrogen, is used as an indication of lactose intolerance in a widely applied breath test. Since milk is the most common dietary source of calcium, people suffering from lactose intolerance are often calcium deficient. Cheese and yogurt contain far less lactose and most lactose-intolerant people can handle these foods. One ounce (twenty-eight grams) of cheddar cheese provides as much calcium as one cup (250 milliliters) of milk, but less than one-tenth as much lactose. A preparation containing the missing enzyme is also now commercially available. Taking Lactaid before drinking milk or ingesting other dairy products leads to the destruction of most of the lactose within twenty-four hours and will prevent the feared side effects.

Now, what about the need for milk to "build strong bones"? The antimilk lobby points out that Asians have a lower incidence of osteoporosis than Westerners even though they consume less dairy. True enough, but they also have a very different overall diet and lifestyle. And then there is the Nurses' Health Study, which found that nurses who drank two or more glasses of milk a day actually broke more bones and had a higher risk of hip fracture. The lead author of this study offers an interesting interpretation. She suggests that women who had the highest risk of osteoporosis drank the most milk, but that it was "too little too late." In any case, when we look at the totality of evidence, an overwhelming number of studies show that bone strength, in the context of the North American diet, improves with calcium intake. And dairy products furnish the best bioavailable source of calcium. When calcium is added to orange juice, for example, bioavailability varies depending on the form of calcium used. Calcium citrate malate is much better absorbed than the combination of tricalcium phosphate and calcium lactate.

How do we know what the ideal calcium intake is? An important clue comes from measuring calcium output in the urine. When the intake is greater than about a gram (1,000 milligrams), the calcium concentration in the urine increases, meaning the body has retained as much as it needs. So it seems that 1,000 milligrams a day is a good ballpark figure. A glass of milk has around 300 milligrams of calcium, a cup of yogurt 400. By comparison, the best vegetable source is broccoli, with about 100 milligrams per cup.

Calcium lactate, calcium gluconate, calcium citrate, and calcium carbonate are all suitable calcium supplements, and are best taken with meals. Calcium citrate is more readily absorbed, but it contains less calcium than calcium carbonate, 24 percent by weight compared with 40 percent. It is also more expensive. Remember that dietary recommendations are always in terms of calcium, which makes up only part of the weight of a supplement. Calcium carbonate is therefore the most efficient source, although it may have a slight constipating effect. As far as the body is concerned, it makes no difference whether the calcium carbonate is manufactured in a laboratory or comes from pearls. Whether one chews on Tums, grazes on the White Cliffs of Dover, or dines on chalk is a question of personal preference. Many calcium supplements now include vitamin D, a good idea.

Milk may not be a miracle food, but it can contribute significantly to a healthy diet. It assuredly is not a poison, as suggested by the likes of People for the Ethical Treatment of Animals. This is the organization that sponsored billboards of former New York City mayor Rudy Giuliani wearing a milk mustache after he had been diagnosed with prostate cancer. Exploiting the dairy industry slogan "Got Milk?" the PETA ads asked, "Got Prostate Cancer?" Here's what I'd ask PETA: "How about the ethical treatment of people?"

PART TWO

MANIPULATING OUR FOOD SUPPLY

FORTIFYING WITH IRON

Iron deficiency is the most common nutritional disorder in the world, affecting up to 25 percent of the population, although only about 5 percent of North Americans. The issue was brought to the public's attention in the 1930s, albeit indirectly, by that nutritional icon we have already encountered, Popeye the Sailor Man. As most everyone knows, when Popeye needed some extra strength he didn't pop steroids, he popped a can of spinach. And why did Elsie Segar, Popeye's creator, use spinach as the sailor's magical energy boost? Because iron can increase energy, and spinach does contain iron. But there are several problems with the Popeye–spinach connection. First of all, increasing iron intake to boost energy works only if there is iron deficiency in the first place, and even then energy will only be restored to normal levels. But there are a couple of other issues as well. Spinach doesn't contain all that much iron in the first place, and what it does contain is not readily absorbed by the body.

By the time that Popeye first appeared in Segar's comic strip in 1929, researchers had established the vital role iron played in nutrition. It was an integral part of hemoglobin, the oxygen-carrying molecule in red blood cells. Lack of iron caused anemia, which was characterized by tiredness, impairment of mental acuity, and even itchiness. (Could this be why we scratch our heads when we think?) Increasing iron intake solved the problem and restored energy.

As the role of iron in health began to be unravelled, it became important to know which foods contained it and how much. There are many chemical ways to determine the iron content of foods. One of the more interesting ones is based on the reaction of iron with thiocyanate to form a red color. The intensity of the color can be used to calculate the amount of iron by comparison to standards. For example, to determine the iron content of spinach, a sample is burned until only an ash remains. Treating a water extract of this with thiocyanate yields a red color, the intensity of which can be analyzed by a colorimeter. It turns out that spinach is not that great an iron source. In the eighteen hundreds, researchers made a mistake and placed a decimal point in the wrong position when they determined the iron content of the vegetable! This mistake was propagated in many texts, leading to Segar's selection of spinach as Popeye's power source. A further problem is that the iron in spinach is not readily available. Naturally occurring oxalate and tannins bind the mineral and prevent it from being absorbed.

So, if spinach isn't a reliable source of iron, what is? Meat contains *heme* iron, which is the most absorbable form, but beans, nuts, and prunes are all good sources. The prime nonmeat source for most people, though, is fortified flour. When nutritional authorities discovered in the mid-1900s that our iron intake was dropping, probably because we were trading in our old iron pots for new-fangled aluminum and stainless-steel cookware, they decided to fortify our flour, and consequently our bread and cereals, with iron.

Fortifying foods with iron is not a novel idea and was described long before people knew anything about biochemistry. In Greek mythology, in order to boost their energy, Jason and the Argonauts drank red wine mixed with iron filings saved from sharpening their swords. This would only have been effective if the sailors suffered from iron-deficiency anemia in the first place, an unlikely event. In the seventeenth century, Dr. Thomas Sydenham, a British physician, routinely treated anemia with iron "steeped in cold Rhenish wine." A century later, doctors were recommending that patients complaining of fatigue eat apples punctured with iron nails (after removing the nails, of course!). This was a surprisingly effective technique because the acidity of the apple helped dissolve some of the iron, and vitamin C, which is found in apples, enhances iron absorption.

Absorption is a very significant problem when dealing with iron fortification. Ferrous sulphate is water soluble and bioavailable, but it can affect the color, taste, and keeping qualities of food. There is not much sense in improving the nutritional properties of a food if people will not eat it. That's why elemental iron powders are used, even though this form of iron is less well absorbed. If you would like to do a neat experiment, mash some Total cereal in a blender and then stir with a magnet. Before long you'll see a coating of tiny iron particles!

Because of the widespread problem of iron deficiency in the world, researchers are constantly looking for improved fortification methods. Much of the developing world consumes whole-grain flour, which is difficult to fortify effectively because of the presence of phytates, which bind iron strongly. Adding vitamin C to increase absorption is a possibility in foods that are not heated, as is the use of chelated iron compounds. In the latter, the iron is complexed either with the amino acid glycine, or with ethylenediaminetetraacetic acid (EDTA), which prevent binding to phytates and enhance absorption. Research into increasing iron intake in developing countries is essential. Although people think of iron-deficiency anemia as mainly causing weakness and lack of energy, the fact is that it can have far more serious consequences, such as complications during pregnancy, increased infant mortality, and impairment of physical and mental development.

A little bit of iron is required, but that doesn't mean more is better. Many Bantu in South Africa, for example, suffer from iron overload because they cook everything in iron pots and drink beer fermented in iron vessels. But of more immediate concern to those in the developed world is increased absorption of iron due to a condition that affects about three in a thousand people, known as hemochromatosis. The symptoms of this disease can be very similar to anemia, and misdiagnosis followed by a recommendation to take iron supplements can be fatal. The appropriate treatment, believe it or not, is bloodlet-ting. Since vitamin C enhances the absorption of iron, supplements of this vitamin can be damaging to people suffering from hemochro-matosis. And unfortunately the only way that the condition can be diagnosed is through a blood test. Most people with the disease do not know they have it until symptoms begin to appear.

There is another issue with iron in the body. In 1992, a Finnish study found that men who had higher levels of ferritin, the body's iron-storage protein, had an increased risk of heart attack. The theory is that iron can catalyze the formation of free radicals, which in turn can damage the lining of arteries and lead to the buildup of plaque. Most subsequent studies have failed to corroborate the connection to heart disease, but some have suggested a link to neurological conditions such as Parkinson's disease. Obviously, we do not want to go overboard with iron intake. Men and older women only need about eight milligrams a day, which is readily available in the North American diet. No need for supplements. Premenopausal women who bleed heavily during menstruation, pregnant women, people on low-calorie diets, and endurance athletes require about eighteen milligrams a day and may benefit from iron supplements, but this has to be discussed with a physician or dietician. Increasing dietary intake from meat, poultry, or fish may be enough. But spinach will not do the job. Let's remember also that increasing iron intake to "gain energy" works only if the lack of energy is due to iron-deficiency anemia. And if this condition is diagnosed, it has to be further in-vestigated because blood loss may be due to an underlying condi-tion such as colon cancer.

Still, this does not mean we should not follow Popeye's advice about spinach. He was wrong about the iron, but spinach turns out to be an excellent source of folic acid, as well as of beta-carotene, both of which contribute to good health. So a spinach salad is a great idea—dressed with a monounsaturated oil that may be instrumen-tal in the reported health benefits of the Mediterranean diet. After all, Popeye loved Olive Oyl.

FLAVORING WITH SALT

The processed-food industry loves salt. Sodium chloride is cheap, allows water to be retained, acts as a preservative, and enhances flavor. As one salt promoter says, "Salt is what makes things taste bad when it isn't in them." True enough. The human craving for salt may partially be explained by our physiological need for sodium. Without it, our nerve cells can't transmit electrical impulses, our muscles can't contract properly, and our body fluids go out of kilter. So it shouldn't come as a surprise that "salty" is one of the basic human tastes. But salt does more than just add saltiness to food; it can also modify the way we perceive the other common tastes, namely sour, bitter, and sweet.

Salt inhibits bitterness and enhances sweetness. That's why you'll find salt in such unlikely foods as chocolate, apple pie, and breakfast cereals. Indeed, studies have shown that consumer acceptance drops dramatically when salt levels in processed foods decline. That would explain the popularity of foods such as dill pickles, hot dogs, sauerkraut, vegetable juices, cottage cheese, olives, canned soups, and pizza, which can have up to a gram of salt per serving. It isn't hard to see how the recommended intake of six grams a day can be exceeded.

Salt was the first seasoning used by our ancestors. They got it by evaporating sea water, or by mining it. The origin of salt deposits in the ground can also be traced back to oceans that no longer exist, so basically all salt is "sea salt." Salt was mined near Salzburg ("City of

Salt") in Austria as early as 6500 BC, and the ancient Romans built large evaporation ponds by the sea to collect salt. In fact, the Romans prized salt so much that soldiers were given a special allowance, known as the "salarium" to purchase it. Our word *salary* derives from this Latin word. Salt was deemed so important that spilling it was thought to foster bad luck by attracting malevolent spirits. Tossing a little salt over the shoulder was the antidote. The grains of salt were supposed to lodge in the spirit's eyes and distract him from the evil he was planning. Spilled salt as an omen of bad things to come was an enduring belief. Leonardo da Vinci's *The Last Supper* clearly shows an overturned salt container in front of Judas, foreshadowing his betrayal of Jesus.

It wasn't only for its taste that salt was so prized; its preservative value was very useful. When the salt concentration outside a bacterial or fungal cell is higher than inside it, water is drawn out of the cell to reduce the outside salt concentration. This process of osmosis dehydrates the cell and eventually destroys it. That's why salt was commonly rubbed into wounds to reduce the chance of bacterial infection. Of course, this also disturbs tissue cells and causes the irritation we associate with "rubbing salt into a wound." Meat used to be preserved by soaking it in a brine solution or by covering the surface with whole grains of salt that were known as "corn," hence the origin of corned beef. Perhaps the most unusual use of salt as a preservative was devised in seventeenth-century England when heads of executed villains were put on public display to deter other criminals. But the heads would quickly rot and attract birds that would strip off the flesh, leaving behind a clean skull, which apparently was less frightening to the populace. The answer to this little problem was to boil the heads in salt water so they would not putrefy.

These rogues were salted after death. But what about the possibility of salting bringing on death? Our body tries to maintain a certain concentration of sodium in the blood. If the amount of sodium rises, more water has to be retained to maintain the same concentration. This means that the blood volume increases and that there is more blood for the heart to pump around the body. The pressure the blood exerts against the walls of the arteries increases, and this can lead to strokes and heart attacks. But if less sodium is taken in, less water is retained, and blood pressure should go down. "Go easy on the salt," is the physician's advice to patients diagnosed with high blood pressure.

Numerous studies have shown that about 50 percent of these patients respond to a low-sodium diet. Why not all of them? Because in reality the situation is more complicated than just a simple balance between sodium and water. Calcium and potassium play important roles as well. In fact, many researchers now believe that increasing potassium and calcium intake is as important as reducing sodium intake for people suffering from high blood pressure. This means more skim milk, more bananas, more oranges.

While no one disputes the low-sodium diet for people with high blood pressure, experts bicker when it comes to making recommendations for the public at large. Some say that asking everyone to try to reduce their salt intake from about nine to six grams a day is not based on science. I think they are wrong. Many people have undiagnosed high blood pressure and would benefit from reduced salt intake. Experiments with chimps have shown that as salt in the diet is increased, blood pressure goes up. Human epidemiological studies have shown the same thing. Populations with low salt intake have lower blood pressure. The Yanomami Indians of Brazil add no salt to their food, and they do not suffer from hypertension—in spite of being surrounded by poisonous snakes, bugs, and researchers constantly wanting to measure their blood pressure. By contrast, North Americans, with their penchant for salty hot dogs, chips, and pizza, are in the midst of a hypertension epidemic. In fact, preserved foods account for 75 percent of our salt intake. A single slice of bread can have as much as half a gram of salt. Whether a reduced-salt diet lowers blood pressure in people who do not have high pressure to start with is irrelevant. Eating fewer salty processed foods automatically leads to a healthier diet. Researchers calculate that a reduction of salt intake from an average of nine grams to six grams would save thousands of lives a year! I'm sure that spokespeople for the influential Salt Institute, an organization that promotes the use of salt, will dispute this claim. But I would take their comments with, well, a grain of salt.

ENHANCING TASTE WITH
MONOSODIUM GLUTAMATE

Chemistry professor Kikunae Ikeda undoubtedly enjoyed eating. Like most Japanese, he was particularly fond of soups prepared with *dashi*, an extract of a seaweed known as *kombu*. While the taste of dashi itself was very mild, it seemed to bring out spectacular flavors when added to other foods. What was responsible for this culinary magic, Ikeda wondered? He decided to find out. Starting with a huge batch of kombu broth, Professor Ikeda managed to isolate a smattering of white crystals that provided an answer to the mystery. When placed on the tongue, these crystals had practically no flavor at all, but they made other foods taste more delicious.

The delectable taste the crystals imparted was different from the classic sweet, sour, bitter, or salty sensations associated with foods. Ikeda gave it the name *umami*, from the Japanese adjective *umai*, meaning "delicious." In 1909, he described his work in a landmark article published in the *Journal of the Chemical Society of Tokyo*, identifying the substance with an umami taste as glutamic acid. The sodium salt of this acid was stable and water soluble, he noted, and was a candidate for commercial exploitation. And so it was that monosodium glutamate, better known as MSG, started its career as a widely used additive to "bring out hidden flavors." Ikeda himself got the ball rolling by patenting and selling MSG as a table condiment under the name Aji-no-moto (essence of taste). Never could the Japanese inventor have imagined that his discovery would end up being embroiled in nutritional controversy, but MSG has been accused of being a chemical culprit instrumental in causing conditions ranging from hypertension, asthma, and depression to attention deficit disorder and "Chinese Restaurant Syndrome." Accusations, however, are not the same as facts.

The commercial success of MSG was almost immediate. Within a few years of Ikeda's discovery, chemists had worked out an economical way to produce glutamic acid through fermentation of beet sugar or corn syrup. Soon canned soups, processed meats, salad dressings, frozen meals, and a host of other foods basked in the glory of the heightened flavor afforded by the addition of MSG. Then in 1968, a bump appeared in the road. And it came in the form of a letter to the *New England Journal of Medicine* (*NEJM*) from Dr. Ho Man Kwok, who thought that his personal adventures in Chinese restaurants were worthy of the medical community's attention. Kwok noted that "I have experienced a strange syndrome whenever I have eaten out in a Chinese restaurant, especially one that served northern Chinese food. The syndrome, which usually begins 15 to 20 minutes after I have eaten the first dish, lasts for about two hours, without hangover effect. The most prominent symptoms are numbness at the back of the neck, gradually radiating to both arms and the back, general weakness and palpitations." The editors of the *NEJM* published the letter under the catchy title "Chinese Restaurant Syndrome," and opened up a can of worms.

Kwok did not blame MSG for his symptoms but did mention it as a possibility. This possibility was turned into a probability by a flurry of letters to the *NEJM* triggered by Kwok's observation. Letters came from physicians and pharmacologists who claimed they had seen clear associations in sensitive people between MSG ingestion and symptoms like those described by Kwok. Furthermore, they added fainting, rapid heartbeat, nausea, and muscular twinges to the list of possible effects. And there were more serious accusations. Dr. John Olney at Washington University reported finding lesions in the brains of mice given MSG in an amount equivalent to that found in a can of soup. Although the evidence was circumspect, food producers decided to eliminate MSG from baby foods.

Then came the rebuttals. In one study, humans were given up to 150 grams of MSG a day for six weeks without any ill effects. (A Chinese meal may contain five grams of MSG at most.) Based on their findings, the researchers concluded that "Chinese Restaurant Syndrome is an anecdote applied to a variety of postprandial [after a meal] illnesses; rigorous and realistic scientific evidence linking the syndrome to MSG could not be found." Others also noted that the subjective symptoms of Chinese Restaurant Syndrome are not consistent, and objective symptoms such as heart rate, blood pressure, and skin temperature are unaltered during an "attack." A number of studies with primates showed no effect upon injection or force-feeding of MSG.

The MSG controversy flared up again in 1992 when the influential CBS program *60 Minutes* aired a piece in which a woman alleged that the failure to recognize MSG as the cause of her stomach aches had led to unnecessary surgery, and a mother claimed that her son's hyperactivity and low grades were due to the additive. Dr. John Olney made an appearance, wrapped in a white lab coat, suggesting, without any supportive evidence, that MSG can cause brain damage in some people. The program was irresponsible, paying only lip service to the massive amount of research since 1968 showing that while there may be isolated, idiosyncratic reactions, MSG was not responsible for the suffering of the people on the program.

Jaded reports about MSG are not restricted to TV programs looking for sensational stories. The National Organization Mobilized to Stop Glutamate (which may have spent all its mental acuity on coming up with a name to produce the unpronounceable acronym NOMSG) routinely releases reams of information about the evils of MSG. These accusations, though, are not supported by the scientific evidence. In 1992, the US Food and Drug Administration (FDA), prompted by the public concern about MSG, asked an independent panel of scientists, the Federation of American Societies for

Experimental Biology (FASEB), to study the issue. A comprehensive report in 1995, based on well-controlled, double-blind studies, concluded that MSG presents no problems at the levels normally used, but a large dose may cause burning sensation, facial pressure, headache, drowsiness, and weakness in a very small percentage of people.

Significantly, though, there was a threshold effect. The symptoms were noted only in people who had ingested over 2.5 grams of MSG at one sitting, which can happen with some Chinese meals. Still, the researchers felt that it was unfair to saddle Chinese food with a pejorative label since many other meals could have equally high levels of glutamate. Instead of "Chinese Restaurant Syndrome," they prefer "MSG Symptom Complex." A Canadian study has further underlined the safety of MSG. An investigation of sixty-one subjects who claimed to be MSG sensitive clearly showed that at less than 2.5 grams there was no difference between MSG and placebo. The average North American consumption of the additive is 0.55 grams per day.

Monosodium glutamate has also been accused of triggering asthma and migraines, with some cases having been well documented. This comes as no surprise: there are hundreds of both naturally occurring and synthetic substances that can have such an effect. Curiously, these associations are hardly ever reported in Asian populations, where MSG consumption is far greater than among North Americans. It is also of interest that parmesan cheese and tomatoes are rich sources of naturally occurring glutamate, yet nobody has claimed to suffer from "Italian Restaurant Syndrome." And nobody has suggested that glutamate, present in mother's milk at about ten times the level found in cow's milk, is a health risk for babies. MSG opponents argue that natural glutamate formed by the breakdown of proteins has a different effect on the body than glutamate used as a food additive. They are not clear on why this should be so, but sometimes mutter about impurities that may be found in the commercially produced substance.

The overall scientific evidence does not support the claim that MSG is harmful. It may, however, serve as a convenient scapegoat for any ill effect after a meal. Surprisingly, around 40 percent of the population report unpleasant symptoms after being tested with any food! Admittedly, some people do show a response to MSG, mostly after eating a significant amount on an empty stomach, but symptoms may vary. In any case, these are transient and benign, and are not reflected by objective measurements, or by blood levels of glutamate. But such observations do not allay the fears of the vocal critics who consider MSG to be a nutritional villain unleashed on the public by an uncaring industry bent on lining its pockets at the expense of human health.

Some Chinese restaurants, concerned about the welfare of their clients (or their cash registers) have taken criticism of MSG to heart and have festooned their windows with signs declaring "No MSG Added," while continuing to load the meals with huge amounts of glutamate from seaweed. The glutamate industry has taken another tack. It plays the "natural" card, using the slogan "Natural, Tasty, Safe" to describe MSG. "Many people believe that MSG is made from chemicals," states a website, "but it is a chemical in the same way that the water we drink and the oxygen we breathe are chemicals."

Of course MSG is a chemical! What else would it be? And there is no shame in that. Everything in the world is made up of chemicals. Whether a chemical is safe or not is not determined by whether it comes from a natural source or not. It is determined by its effects on the body as evaluated by careful testing. The reason that MSG has a commendable safety profile is due to the results of studies, not due to its "natural" origin. It would be more appropriate for the glutamate producers to refer to the fact that no official government or academic body has ever issued any warning about consuming monosodium glutamate.

SWEETENING WITH SUGAR AND
HIGH-FRUCTOSE CORN SYRUP

We love sweet things. Sugar-laden cakes, cookies, ice cream, and soft drinks are standard North American fare. There is sugar in our cereal, our bread, and even in our ketchup. We add it to our coffee and tea. All together, on average, we down about fifty teaspoons (or 200 grams) of sugar a day, a truly astounding amount. A single can of soda can harbor as much as ten teaspoons (or forty grams) of sugar. In this case, by *sugar* we mean not only sucrose, the refined white crystals that are produced from sugar cane or sugar beets; we also include high-fructose corn syrup, which is increasingly replacing cane sugar as the prime sweetener in processed foods. Why? Because it is cheaper to produce.

High-fructose corn syrup is made from glucose, which is readily available from corn starch. And given that in the United States corn production often receives government subsidies, there is plenty of the stuff around. Bacterial enzymes are used to break cornstarch down to glucose, which itself can be used as a sweetener. Glucose, however, is only about 70 percent as sweet as sucrose, which is a problem. This is where another enzyme enters the picture. Glucose isomerase, from a special strain of *Streptomyces murinus*, can convert glucose into fructose, which is 30 percent sweeter than sucrose. Furthermore, fructose is more water soluble than glucose, making possible the production of a stable syrup with a roughly 55 percent fructose content. This "high-fructose corn syrup" (HFCS) is cheap, and is more easily blended into soft drinks and foods than sucrose.

Obviously, the sugarcane industry was not happy with the emerging competition from HFCS, but what about any impact on the consumer? Are there any health implications? On first glance this seems unlikely. Sucrose is a disaccharide, composed of a molecule of glucose joined to one of fructose. Indeed, when we ingest sucrose, most of it is broken down into glucose and fructose, so that sucrose can actually be thought of as a 50 percent fructose product. Is it possible that the extra 5 percent of fructose in HFCS can make a difference in the way the body handles the sweetener? Maybe so.

Fructose digestion, absorption, and metabolism differ from those of glucose. For example, glucose is more adept at stimulating the production of leptin, a hormone that curbs the appetite. This is a consequence of the fact that glucose triggers the release of insulin, which in turn leads to the production of leptin. Fructose, by contrast, does not prompt any insulin secretion by the pancreas, an advantage for diabetics but not for people trying to control their weight. A further problem is that leptin reduces the rate at which a major hunger hormone, ghrelin, is released by cells in the stomach. So, reduced leptin production means more hunger pangs. And to complicate matters even more, fructose is more readily converted into fat inside cells than is glucose. Fructose malabsorption is another issue. Many people may experience gas, abdominal distress, and loose stools as a result of increased fructose intake, but they may never relate the symptoms to the presence of HFCS in their diet.

It may seem odd that fructose, the sugar commonly found in fruits, should be linked to such problems. After all, we are constantly being urged to eat more fruit. But let's remember that the fructose in fruits is accompanied by many other healthy nutrients. This cannot be said for the presence of HFCS in soft drinks. Let's compare an apple to a serving of soda. The apple has about ten grams of fructose, the soda about twenty-five. And fiber in the apple slows the absorption of glucose, resulting in a smaller effect on metabolism. Of course, the apple also contains a variety of antioxidants absent in the soft drink.

Neither sucrose nor fructose is a poison, as some undereducated nutritional gurus would have us believe. The problem is overconsumption. According to the World Health Organization, our intake of added sugars in food or beverages should contribute no more than 10 percent to our total daily calorie intake. We consume way more than that, the extra calories contributing to the obesity epidemic in North America—and contributing to cavities as well.

Bacteria that live in our mouth just love sugar. When they metabolize it, they produce acids that eat away the enamel and cause cavities. But these bacteria also feast on starches that they can break down into glucose, which subsequently is also converted into acids. So what is worse in terms of causing cavities, jelly beans or chips? The sugar in jelly beans is soluble and gets washed away by saliva, but the complex carbohydrates in the chips are insoluble and stick between the teeth, providing food for the acid-producing bacteria. Similarly, sugary soft drinks don't spend a lot of time in contact with your teeth, but constantly sucking on candies will lead to cavities.

So much for the facts about sugar. What about the myths? Perhaps the biggest one surrounding sugar is its reputed effect on behavior, especially in children. How often have you heard a parent complain about a child "bouncing off the walls" after a sugary snack? The connection was first suggested in 1922, but it was in the 1970s that the lay literature took up the cause, dubbing the questionable condition *functional reactive hypoglycemia*. Desperately searching to find a reason for bad behavior, parents and teachers began to see links between eating sugar and hyperactivity. But wait a minute. Could it not be that children eat more sugary foods during activities that are conducive to adverse behavior, such as birthday parties? And that sugar does not cause the problem?

Studies tell us that this is indeed the case. When researchers compare children given sugar with children given a placebo, they find that not only does sugar not cause hyperactivity, it can have a calming effect! There actually is a scientific rationale for this finding. Sugar intake elevates levels of a chemical in the brain called serotonin, which has a calming effect. Why do the results of controlled studies differ so much from the impressions of parents? It may be due to parents' expectations as shown by an intriguing experiment carried out by a British television series. *The Truth About Food* decided to put the sugar–hyperactivity link to a more or less scientific test. The producers organized two parties for children. When parents dropped their kids off for the first one, they saw tables loaded with sugary snacks. But as soon as they left, the junk food was replaced by healthy snacks and the children were entertained with some high-energy music and activity. Two weeks later the same children were invited to a party, this time with a sedate storyteller providing the entertainment. A feast of healthy snacks was laid out for the parents to see, but was quickly replaced by cakes, cookies, and soft drinks after they left. After each party, parents were asked to evaluate their child's behavior, and there was consensus that the first party had made them hyperactive. This came as no surprise to the parents, having seen the sugary snacks to which their kids had been supposedly treated. Had the ruse not been revealed, the experiment would have reinforced the parents' conviction that sugar causes hyperactivity. In truth, the ill behavior after the party was caused by the excitement of the event, the frantic music, and the running around. The second party was a calm affair and the children were delivered to their parents in a peaceful state despite the high sugar load.

Recently, though, the advocates of the sugar–hyperactiviy link found some ammunition in a Norwegian study that examined the dietary habits of some 5,000 teenagers. Researchers found a significant association between sugary beverage intake and hyperactivity, as well as a complex association with other mental problems. The worst hyperactivity problems were seen in teenagers who drank more than four soft drinks a day. This is not a very unusual amount; about 10 percent of those surveyed drank at least four such beverages a day. Curiously, those who drank no soft drinks were more likely to have mental problems. In any case, associations such as these cannot prove cause and effect, and it is possible that hyperactive teenagers are more likely to consume soft drinks. Whether the hyperactivity link is real or not, cutting back on sucrose (table sugar) is sound scientific advice. Flooding the bloodstream with sucrose causes a burst of insulin to be released that can then drop the sugar level in the blood quickly, sometimes to levels below normal. This may result in muddied thinking and poor classroom performance. But other nutritional factors, such as the type of fats in the diet, likely play a bigger role in determining children's behavior. Fats are an integral component of cell membranes and determine the fluidity of these membranes, which in turn affects the way cells communicate with each other through chemicals called neurotransmitters.

With the advent of processed foods, our pattern of fat consumption has changed. Our intake of trans fats from processed foods and omega-6 fats found in corn and soy oil has increased, while our intake of omega-3 fats found in fish and vegetables has decreased. This decrease may affect behavior, and some studies have shown improvement in children's behavior when their diet is supplemented with omega-3 fats. There is also some evidence that gluten in wheat or casein in milk can have an adverse effect on behavior, as can certain food colorants. While these findings are debatable, there is no doubt that a diet with fewer processed foods and less sugar is preferable for all sorts of reasons. So give your kids apples and carrot sticks instead of cakes and ice cream at the next party, but if you want good behavior, hire a cellist instead of a clown.

CUTTING CALORIES WITH "NON-NUTRITIONAL" NATURAL SWEETENERS

There's no question that many people look in the mirror and don't like what they see. Those extra pounds, often the result of eating too many sweets, are certainly not flattering. But cutting sugar out of our diet is a difficult task. Its taste is just too seductive! And so the chase is on to find a way to keep the sweet taste but eliminate the calories. Several possibilities come to mind. The most obvious idea is to find a substance that is far sweeter than sugar, meaning that only tiny amounts are needed for a sweetening effect. Alternatively, we can look for substances that provide a sweet taste, but are poorly absorbed by the body. If they are not absorbed, they cannot provide calories.

We can search for such substances in the natural world, or we can look to the chemist's ingenuity to come up with synthetic sugar substitutes. But there's more than science at stake here. Potential profits in the sweetener market are huge, so competition is fierce. The sugar industry aims to maintain its stranglehold on our taste buds and vigorously attacks competitors whenever it feels threatened. "Non-nutritive" sweetener producers fight back enthusiastically against Big Sugar, but have to reserve some energy to fight among themselves for market share. And to complicate the issue further, special-interest groups with various agendas get embroiled in the battle, often claiming that artificial sweeteners are foul substances cooked up by a wicked industry that cares only about profits. Unfortunately, in the sweetener wars, science often takes a back seat to posturing.

Let's start our journey through the sweetener quagmire with stevia, a widely promoted "natural" sweetener. Is there really a conspiracy to keep this "safe, noncaloric, natural sugar substitute" off the market? That is just what promoters of various stevia preparations claim. And who is behind the conspiracy? Sugar producers and manufacturers of artificial sweeteners, of course, who worry that their profits will nosedive if stevia is allowed as a food additive. Malarkey, according to the US Food and Drug Administration and Health Canada. Stevia is not allowed as a food additive for the simple reason that there are outstanding questions about its safety.

Well, the Guaraní natives of Paraguay don't think there is a problem with stevia. They have been sweetening their traditional brew of yerba maté with it for centuries. *Stevia rebaudiana* is a shrub native to South America that contains a number of naturally occurring compounds with a remarkably sweet taste. Stevioside and the related rebaudiosides are hundreds of times sweeter than sugar, meaning that very little of these compounds is needed to achieve a sweet taste. In Japan, purified stevioside is widely used in foods and beverages as an additive, including in Diet Coke and in sugar-free gums. Ditto in Paraguay and Brazil. So why is stevia deemed safe as an additive in these countries, but not in Canada or the United States? Our authorities claim that it is because we have a very stringent regulatory system and that stevia producers have not furnished the required documentation to establish its safety.

If stevia producers want to market their product as a food additive, they have to meet the same standards as the manufacturers of any other artificial sweetener. This, according to both the Canadian and American governments, has not happened, and questions about the safety of stevia have not been properly answered. Government scientists point to studies in which male rats fed high doses of stevia for twenty-two months showed a reduction in sperm production and an increase in cell proliferation in the testicles. In female rats, large doses of steviol, a breakdown product of stevioside, reduced the number and weight of offspring.

Obviously, the Japanese, Chinese, Koreans, and South Americans don't attach much importance to these studies, judging by their approval of stevioside as a food additive. And apparently they haven't seen any adverse reactions in humans. But the fact is that the amount of artificially sweetened products consumed in these countries is small. That would hardly be the case in North America if stevia were approved as an additive. People who are leery of aspartame and saccharin might hop on the stevia bandwagon and in all likelihood indulge to a far greater extent than the Japanese, who presently happen to be the biggest users of stevia.

Both Canada and the United States are considering approval of stevia as a food additive, and a decision is imminent. Stevia preparations can already be sold as dietary supplements, which are regulated differently from food additives. Tablets containing the crushed leaves, liquid extracts of the leaves, as well as purified stevioside are all available. Judging by the historical evidence, use of these products in moderate doses is likely to be safe. As far as frequent use goes, though, nobody can say. It isn't any conspiracy that is keeping stevia from being marketed as a food additive; it is lack of submitted evidence of safety.

I must admit that looking into the stevia situation whetted my appetite for the sweetener. And what better way to try it than by adding it to a traditional brew of yerba maté? After all, I had seen plenty of ads on the Web boasting of this potion's "powerful rejuvenating effects." Well, who couldn't use a little rejuvenation? It turns out that yerba maté is a tea brewed from the dried leaves of the *Ilex paraguariensis* plant, a small shrub that, like stevia, grows in Paraguay, Brazil, and other South American countries. The tea is also sometimes known as Paraguay tea and has a reputation for boosting energy levels and even intelligence. In Europe yerba maté extract is often used for weight loss, although there is no scientific evidence to show that the plant boosts metabolism or acts as an appetite suppressant. But what about the rejuvenation claims?

An analysis of extracts taken from the maté plant reveals the presence of a couple of hundred compounds, as one would expect for any plant material. There are vitamins and minerals and the usual array of antioxidants, but no magical ingredients are apparent. Any stimulation from the beverage can probably be ascribed to caffeine, although yerba maté contains less than coffee or other teas. Claims about yerba maté being "nature's most perfect beverage" or "the beverage of the gods" are just hot air. And speaking of hot, that's how yerba maté is traditionally consumed. That can be a problem. Drinking maté tea has been linked to esophageal cancer in South America, where the beverage is consumed at extremely high temperatures.

Maybe I didn't brew my yerba maté properly, but to me it tasted like a mix of bad coffee, green tea, and fermenting grass. I felt more nauseated than rejuvenated. Incidentally, the Guaraní natives traditionally drink yerba maté out of a bull's horn, which seems appropriate, given some of the outlandish claims that are made on behalf of the beverage. It surely would have tasted a lot worse, though, had I not added the stevia leaves. While we may not know enough about the long-term effects of high-dose stevia consumption, I can certainly attest to the amazing sweetening power of the small doses of steviosides found in the leaves of this remarkable plant.

But you don't need such an intense sweetening effect if you can find a sweetener that is not readily absorbed by the body. In fact, besides reducing the worry about calories, such sweeteners have an advantage. They can add "bulk" to a product. Consider chocolate bars as an example. As far as sweetening power goes, sugar in chocolate can be replaced by artificial sweeteners such as aspartame, acesulfame-K, or sucralose. All of these are hundreds of times sweeter than sugar, meaning that only small amounts are necessary. But sugar not only sweetens, it also provides bulk and an appealing texture to chocolate. You just cannot make an appealing bar by simply replacing sugar with artificial sweeteners. This is where the sugar alcohols, or as they are commonly known, the *polyols* come in.

Polyols are carbohydrates that provide sweetness but are metabolized by the body in a different way from sugar. They occur naturally in various fruits and vegetables but can also be readily synthe-sized from naturally occurring sugars. Lactitol, for example, the polyol used in many chocolate products, is made by reacting the milk sugar lactose with hydrogen gas. Similarly, glucose can be converted to sorbitol, maltose to maltitol, and mannose to mannitol, all polyols that are used in a variety of sugar-free gums, ice creams, candies, and cookies. Polyols are effective sugar replacers because they can basically substitute for an equal amount of sugar. Since they are somewhat less sweet than sugar, an artificial sweetener such as sucralose is also commonly added to boost sweetness. But what is the point of replacing one carbohydrate with another?

Sucrose, or table sugar, is composed of a molecule of glucose joined to one of fructose. During digestion in the stomach and small intes-tine, the link is broken and the glucose and fructose are absorbed into the bloodstream, ready to serve as a source of energy. A gram of su-crose "contains" four calories, meaning that we have to "spend" four calories worth of exercise to use up the sugar. If we don't do that, the excess sugar can be converted into fat, ready to be stored by the body. Now let's turn to lactitol. This compound resists absorption into the bloodstream from the stomach and small intestine. While some is slowly absorbed, much of the lactitol moves down through the small intestine and migrates into the colon. Here it encounters a variety of bacteria. Some of these microbes consider lactitol to be a tasty morsel and make a meal of it. Unfortunately, these bacteria are quite flatu-lent and produce gases as they feast on the lactitol. Buildup of the effluvia can cause bloating and cramping. Furthermore, the body tries to eliminate unabsorbed lactitol, sometimes resulting in an unpleas-ant laxative effect. So, where's the upside?

First of all, a nutrient that isn't absorbed by the body cannot con-tribute calories. Lactitol, which is only partially absorbed, provides two calories per gram, as compared with the four supplied by sugar. Basically, this means that only half as much activity is required to "burn up" the calories in a gram of lactitol when compared with a gram of sugar. Remember, though, that most of the calories in a chocolate bar come not from the sugar but from the fat in the cocoa butter used to make the chocolate, and sugar-free chocolates contain no less fat than regular chocolates. Replacing sugar with lactitol results in only about a 20 percent saving in calories, not a particularly significant amount. An interesting potential benefit, though, lies in lactitol's ability to serve as a *prebiotic*. At a daily dose of five to ten grams, it encourages the growth of beneficial bacteria in the colon at the expense of disease-causing bacteria. Some of the organic acids, metabolites of the beneficial bacteria, have potential anticancer properties. Then there is the fact that while bacteria in our colon like lactitol, those in our mouth do not, and therefore they do not produce the cavity-causing acids they do when they come into contact with sugar.

Now, what about the portion of lactitol that is absorbed into the bloodstream? Unlike most carbohydrates, it is not readily converted to glucose and is therefore less likely to trigger an insulin response. This means that diabetics who have to count carbohydrate exchanges can eat more of the sugar-free chocolate than regular chocolate for the same exchange value. Whether or not one wants to eat more of this chocolate is another question. The party line is that lactitol and other sugar alcohols, when consumed in moderation, should produce no undesirable side effects. But the fact is that in some people, even small doses of sugar alcohols can cause temporary bloating, diarrhea, and impressive flatulence.

Sugar alcohols then have some commercial appeal but are not the ideal "natural" sweeteners. Wouldn't it be great if we could come up with a natural sugar substitute that tastes like sugar, can be used in bulk like sugar, browns with heat like sugar, doesn't rot your teeth like sugar, has a fraction of the calories of sugar, and may even be good for you? You are probably thinking, Dream on! Well, this dream may actually become reality. Tagatose just may be one of those products that lives up to its advance billing. Its appeal comes partly from the fact that it isn't exactly a substitute for sugar, it *is* a sugar.

Sugar substituting for sugar? That probably sounds confusing. But the term *sugar* has a different meaning to the chemist than to the layperson. To most people, sugar is sucrose, the sweet crystals isolated from sugar cane or sugar beets. To the chemist, however, *sugar* is a term that describes a family of carbohydrates that have very similar chemical structures and taste sweet. So sucrose, lactose, glucose, and fructose are all examples of sugars. And so is tagatose, the star of our story.

The sweetness of a substance is a reflection of its molecular structure. Molecules with a particular shape fit into sweetness receptors on our taste buds, much like a key fits a lock. This interaction then stimulates nerves to send the message of "sweetness" to the brain. Sucrose, or table sugar, is an excellent fit. Fructose, commonly known as fruit sugar, fits even better and has a sweeter taste. But the problem is that after stimulating our taste buds, these sugars are absorbed into the bloodstream, and if they are not burned for energy, they're converted to fat and contribute to weight gain. The rate of absorption of sugars through the intestinal wall is also a function of molecular structure. This then suggests the possibility of synthesizing sugars with molecular structures that fit sweetness receptors, but that are not well absorbed from the gut.

Everything in the world, save perhaps a vampire, has a mirror image. Molecules are no exception. But mirror images are curious things. Imagine placing a ping-pong ball in front of a mirror. If you could somehow pick up its image from behind the mirror, you would find it to be identical with and exactly superimposable on the original. But now imagine that you hold your left hand in front of the mirror. The reflection you see is a right hand. Were you able to pick up this mirror-image hand, you would find that it is not identical to the original. What's the difference between the ping-pong ball and your hand? The ball is symmetrical, the hand is not. Basically, any object that is not symmetrical will have a nonidentical mirror image. Sugars are not symmetrical and can therefore exist in "left-handed" and "right-handed" forms. But for a few exceptions, the sugars that are found in nature all have the same "handedness" and are referred to as D-sugars.

One idea then is to synthesize in the laboratory a mirror-image sugar, or L-sugar, and hope that it retains D-sugar's sweetness but not its rapid absorption properties. It turns out that this is indeed the case for glucose and its sweeter cousin, fructose. Unfortunately, so far these L-sugars have defied attempts at a viable commercial synthesis. An astute researcher at the Spherix Corporation did note, however, that tagatose, a sugar found in small amounts in dairy products, had a very similar molecular structure to L-fructose. This sugar, almost as sweet as sucrose, had been known for a long time, having been initially isolated from the gummy resin of a type of evergreen tree, but nobody had previously examined its absorption properties. Experiments first with rats and then with humans showed that the absorption of tagatose was very inefficient; most of a dose passed through to the large intestine. This means that the effective caloric content of tagatose is much less than that of sugar, only about 1.5 calories per gram as opposed to four calories per gram. And there is even more good news. In the large intestine, bacteria break tagatose down into short-chain fatty acids, which have been linked with protection against colon cancer. There may be other benefits as well. Tagatose taken with meals has been found to improve blood glucose levels in type-2 diabetics. Extensive testing in humans has revealed no significant adverse reactions, but in some cases excessive consumption may cause mild intestinal discomfort and perhaps loose stools.

The US Food and Drug Administration is convinced of the safety of tagatose and has approved its use in food. Methods have already been worked out to produce this sugar economically from lactose (milk sugar), which is readily available from whey. No milk protein or lactose remain in the final product, so consumers who suffer from milk allergy or lactose intolerance can safely consume tagatose. Clearly, tagatose by itself will not solve the obesity problem in North America. You still can't have your cake and eat it too, but if it's sweetened with tagatose, at least there may be less of you!

CUTTING CALORIES WITH
"NON-NUTRITIONAL" ARTIFICIAL SWEETENERS

Given their lucrative potential, one would think artificial sweeteners were developed by clever chemists working for companies dedicated to the problem of finding a replacement for sugar. This, however, is not the case. Most of the artificial sweeteners on the market were discovered accidentally, often as a result of sloppy laboratory activities. The scientists involved, though, were clever enough to realize that an important discovery had been made.

The first artificial sweetener ever to make it to market was saccharin. And like the others that would follow in its footsteps, it was shrouded in controversy right from its beginnings. Constantine Fahlberg, a German chemist, had come to the United States to work with Professor Ira Remsen, a leading light at Johns Hopkins University in Baltimore. The project Fahlberg was assigned was not particularly exciting. He was asked to study the oxidation of certain coal-tar derivatives known as toluene sulfonamides. It seems the German was a pretty sloppy chemist who usually didn't even bother to wash his hands after leaving the laboratory. That sloppiness, though, turned out to be his stroke of luck.

One day, at dinner, Fahlberg noted that a slice of bread he had picked up tasted unusually sweet and quickly traced the sweetness to a substance he had been handling in the laboratory. He brought this to the attention of Remsen, and in 1880 the two scientists published the finding in the *American Chemical Journal*, noting that the new compound was hundreds of times sweeter than sugar. Remsen looked upon this as a mere curiosity, but Fahlberg immediately saw the potential for commercial exploitation. He knew that sugar prices fluctuated greatly and that a low-cost sweetening agent would be most welcome. Dieters, Fahlberg thought, would also find the new product appealing. Fahlberg coined the term *saccharin* for his discovery, after the Latin word for sugar, and secretly patented the process for making it. Within a few years saccharin became the world's first commercial non-nutritive sweetener and made Fahlberg a wealthy man.

Remsen did not resent the fact that neither he nor Johns Hopkins University ever made a dime from saccharin. He was a pure scientist at heart and did not much care whether his research turned out to be financially profitable. But he did develop an intense dislike for Fahlberg, who by all accounts tried to take sole credit for the discovery. "Fahlberg is a scoundrel," Remsen often said, "and it nauseates me to hear my name mentioned in the same breath with him!" But like it or not, the importance of the saccharin discovery has forever linked the names of the two men. First, the commercial production of saccharin represents the earliest example of "technology transfer" from university research to the marketplace. But even more importantly, saccharin introduced the concept of a non-nutritive sweetener, an idea that has remained mired in controversy.

Saccharin first went into commercial production in Germany, where Fahlberg had taken out a patent. It wasn't until 1902 that John Francis Queeny, a former purchasing agent for a drug company in St. Louis, decided to take a chance on manufacturing saccharin in the United States. Here the sweetener was not burdened by any of the legal problems that were arising in Europe. He borrowed $1,500 and founded a company that at first had only two employees, himself and his wife. Queeny decided to give the company his wife's maiden name and Monsanto was born. At first, the company's only product was saccharin, but it quickly diversified to become one of the largest chemical companies in the world.

The sweetener met its first enemy in the person of Dr. Harvey W. Wiley, who in 1883 had been made chief of the Department of

Agriculture's Bureau of Chemistry. Wiley had become concerned about the unregulated use of food additives, an issue to which he had become sensitized in his days as a professor of chemistry at Purdue University. He became a zealot for food safety, and saccharin was caught up in the net he cast to catch chemical culprits. He vigorously attacked saccharin as a "coal-tar by-product totally devoid of food value and extremely injurious to health." Unfortunately for Wiley, President Theodore Roosevelt had been prescribed the sweetener by his physician and loved the stuff. "Anyone who says saccharin is injurious to health is an idiot," Roosevelt proclaimed, and decided to curtail Wiley's authority. The president established a "referee board of scientists," ironically with Ira Remsen as its head, to scrutinize Wiley's recommendations. The board found saccharin to be safe but suggested its use be limited to easing the hardship of diabetics. That suggestion had no legal bearing and was soon forgotten in the face of massive industry maneuvering to satisfy the public's demand for non-nutritive sweeteners.

In 1977, saccharin ran into a problem again when a Canadian study suggested an increased incidence of bladder cancer in male rats fed the equivelant of 800 diet drinks a day, but only if their mothers had also received the same dose. Based on this study, ridiculed by saccharin promoters as irrelevant for humans, Canada banned saccharin as a food additive but allowed its use as a tabletop sweetener to continue. The US Food and Drug Administration (the successor to Wiley's bureau) also proposed a ban, but a massive public outcry caused Congress to put a moratorium on removing saccharin from the marketplace, pending further studies. Its continued use as an additive was allowed, but a warning label stating that saccharin "has been determined to cause cancer in laboratory animals" was mandated on the familiar little pink packets.

Subsequent research failed to clear saccharin of all blame as a carcinogen, but human epidemiological studies have shown that if there is any risk, it is very small. Finally, in 2000 the American government removed saccharin from its official list of human carcinogens and President Clinton signed a bill eliminating the requirement for a warning label on the product. Canada still does not allow saccharin as an additive.

The situation is quite different for sodium cyclamate, which it seems may cause cancer in the United States but not in Canada or in some fifty-five other countries around the world. This artificial sweetener was banned by the FDA in 1969 but enjoys brisk sales elsewhere, demonstrating that different countries come to different conclusions based on the same scientific evidence. How can this be? Either the evidence is not conclusive, or perhaps factors other than pure science are involved.

The sweetening power of sodium cyclamate was discovered in 1937 by Michael Sveda, a graduate student at the University of Illinois, who was researching fever-reducing drugs. Although today it would be unthinkable, Sveda routinely smoked in the laboratory. One day, as he brushed some tobacco remnants from his lips, he noted an unusual taste. As he later commented, "It was sweet enough to arouse my curiosity." At the time the only sugar substitute available was saccharin, but it had a bitter aftertaste and the market was ripe for a better sweetener. Sveda, recognizing the potential of his discovery, applied for a patent, which was eventually acquired by Abbott Laboratories. A decade of research on the product's safety led to FDA approval in 1950. By this time obesity was becoming an issue and Abbott began to advertise cyclamate as a low-calorie sweetener rather than as a cheap alternative to sugar. Cyclamate was only thirty times sweeter than sugar, far less than the sweetening power of saccharin. But mixing ten parts of cyclamate with one part of saccharin yielded a product with no aftertaste. "Sweet'N Low" quickly conquered the artificial sweetener market. By the late 1960s, some twenty-one million pounds (955,000 kilograms) of cyclamate were being consumed by the American public annually in products ranging from soft drinks to salad dressings.

Humans were not the only ones guzzling cyclamates. Rats were doing it too. Although cyclamate had been FDA approved, research about its properties continued and chinks in the armor began to appear. By 1966, scientists had discovered that bacteria in the gut could convert cyclamate into cyclohexylamine, a substance that had toxic potential. This triggered various rat feeding studies as well as experiments that involved injecting chicks with cyclamate. In one study, a mixture of saccharin and cyclamate in amounts equivalent to a human drinking 350 cans of diet soda a day caused bladder tumors in rats, but the most dramatic result was noted in 1969 when FDA scientist Jacqueline Verrett appeared on NBC *Nightly News* with pictures of malformed baby chicks that had been injected with cyclamate. "More dangerous than thalidomide," she declared, and the FDA promptly banned cyclamates in 1970.

The Canadian government did not see the great risk but did restrict cyclamates to use as tabletop sweeteners. More than thirty studies since that time, some on primates, have failed to corroborate the supposed danger of cyclamate, but it remains banned in the United States, and some observers argue that this is the result of effective lobbying by the sugar industry. Unfortunately, there are no human epidemiological studies available because virtually no one uses only cyclamate as a sweetener; people tend to use a combination of products. Some researchers contend that while cyclamate may not be a carcinogen, it can increase the cancer-causing potential of other substances, but there is no evidence for this belief. In any case, the cyclamate sprinkled into coffee has less carcinogenic potential than the naturally occurring carcinogens like benzene and furfural found in the beverage.

As in the case of cyclamate and saccharin, the sweetening power of acesulfame potassium (acesulfame K) was discovered through sloppy laboratory technique. This time it was Karl Clauss, synthesizing some novel molecules at the Hoechst Chemical Company in Germany, who in 1967 licked his finger and noted a sweet taste. He immediately recognized the market potential of his finding, but about twenty years of testing were needed before acesulfame won approval as an artificial sweetener. The compound is about 200 times sweeter than sugar, and unlike aspartame, does not lose its sweetness when heated. Given that about 95 percent of acesulfame consumed is excreted unchanged in the urine, and that American, Canadian, and European agencies have carefully reviewed safety assessment studies and have found no problems, you might think that there would be no opposition to the use of this sweetener. But you would be wrong. No matter what is introduced into the marketplace, there will be some individuals and some organizations that make accusations of inadequate testing and maintain that the industry is playing Russian roulette with the health of consumers.

In the case of acesulfame, the claim is that the tests were carried out by Hoechst, a company that had a vested interest in the product. The rat studies were not long enough, critics say, or the dosages used were too low, or an increase in breast tumors in female rats was ignored. In fact, authorities have examined all of these claims and found them lacking in substance. The current scientific opinion is that a ballpark figure of ten to fifteen milligrams per kilogram of body weight is an acceptable daily intake and presents no problems. And how does that relate to consumption? Well, a 220-milliliter can of Coca-Cola Zero has thirty milligrams of acesulfame, so an average adult could drink at least twenty such cans a day and still be well below the acceptable daily intake. Why is there acesulfame in Coca-Cola Zero? Unlike Diet Coke, which is sweetened with aspartame, Zero uses a blend of aspartame and acesulfame. This highlights one of the interesting features of acesulfame, namely that in combination with other sweeteners it masks unpleasant aftertastes, while synergistically enhancing sweetness. A combination of acesulfame and aspartame is about 300 times sweeter than sugar, an improvement over the sweetening ability of the individual compounds.

Aspartame is the most widely used and perhaps the most controversial artificial sweetener. It was back in 1965 that G. D. Searle Company chemist Jim Schlatter tasted the compound he had just synthesized in his lab. Schlatter had not set out to find a sugar substitute; he was carrying out research on gastric ulcers. He knew that entry of food into the stomach stimulates the secretion of gastrin, a hormone that triggers the production of gastric acid. At the time, the common belief was that ulcers were caused by excess stomach acid, and Schlatter was interested in finding a drug that could inactivate gastrin. In the course of this research he synthesized some model compounds that incorporated certain features of the hormone. One day, after licking his finger to pick up a sheet of paper, he noticed a sweet taste, which he eventually traced to the aspartylphenylalanine methyl ester he had just synthesized in the lab. Little did Schlatter dream that within twenty years his discovery would be netting the company $1 billion of profit a year! And he most certainly never imagined that his sweet crystals would become embroiled in a bitter scientific controversy.

We'd best try to make sense of the controversy by discussing the facts about aspartame. The sweetener is commonly labelled as "noncaloric," although that terminology is technically not accurate. Aspartame is broken down in the digestive tract into its components, namely aspartic acid, phenylalanine, and methanol, which are absorbed and metabolized. Collectively, they contribute about four calories per gram, but since the substance is about 180 times sweeter than sugar, very little needs to be used in foods and beverages to achieve a satisfactory degree of sweetness. So the calorie contribution is essentially irrelevant. Aspartame cannot be used in cooked or baked foods because it breaks down into its components upon exposure to heat and loses its sweetening power.

Diet drinks normally contain about 60 milligrams of aspartame per 100 milliliters, which translates to roughly 200 milligrams per serving. To put this into perspective, we need to introduce the concept of *Acceptable Daily Intake* (ADI), which the US Food and Drug Administration uses to describe an intake level that, if maintained each day throughout a person's lifetime, would be considered safe. The ADI for aspartame is fifty milligrams per kilogram of body weight. The actual average daily intake is less than 2 percent of this, and even the heaviest aspartame consumers ingest only about 16 percent of the ADI. To reach the ADI, an adult would have to drink twenty 12-ounce (355-milliliter) soft drinks, and a child would have to drink seven. An adult would have to consume ninety-seven packets of tabletop sweetener. Industry figures show that 99 percent of aspartame users consume less than thirty-four milligrams per kilogram of body weight per day. The average consumption is about 500 milligrams per day. This is a lot less than 3,500 milligrams, which is the ADI for someone weighing seventy kilograms (154 pounds).

Without a doubt, the three breakdown products of aspartame are all toxic in high doses. Phenylalanine is an essential amino acid that must be included in the diet for normal growth and tissue maintenance, but sustained high blood levels of phenylalanine can lead to brain damage. This is of major concern to roughly one out of 20,000 children who are born with an inherited condition called phenyl-ketonurea or PKU. These children cannot metabolize phenylalanine properly, so it builds up to dangerous levels in their brains. The condition necessitates a severe curtailment of phenylalanine intake at least for the first six years of life. This means that aspartame, due to its phenylalanine content, is not suitable for PKU sufferers, and a warning to that effect appears on products in which it is an ingredient. Neotame, a descendant of aspartame, avoids this problem. This slightly altered molecule retains aspartame's intense sweetness but does not yield phenylalanine in the body.

In the general population, phenylalanine levels in the blood after aspartame ingestion are in the same range as after eating any protein-containing food. Even at abusive amounts, equivalent to a child swallowing 100 sweetener tablets, levels do not rise above those considered to be safe in children afflicted with PKU. The effects of aspartic acid, another aspartame breakdown product, have also been rigorously examined. Administration of extremely large amounts to nonhuman primates produced no damage even though blood levels were greatly elevated. In humans, even high doses are quickly eliminated. Most significantly, aspartic acid levels in the blood are not increased after eating foods containing aspartame, or when drinking sweetened beverages even at the rate of three drinks in four hours.

But what about issues other than acute toxicity or worries for PKU sufferers? Why is aspartame one of the most contentious food additives on the market in spite of the stringent regulatory processes in place? Much of the worry is fuelled by a plethora of websites dedicated to demonizing aspartame as a cause of cancer, heart disease, depression, headaches, seizures, visual problems, multiple sclerosis, Parkinson's disease, hair loss, and even breast enlargement in men. The biochemistry involved in such conditions is diverse, and it would take a remarkable substance indeed to cause all these problems—not one that is metabolized to two common amino acids and a small amount of methanol!

Perhaps the most vehement accusations levelled at aspartame have involved its potential to liberate methanol. In large doses methanol can lead to blindness and even to death, and alarmists have referred to the methanol released from aspartame as an unsafe substance. But there are no safe substances, only safe doses. A liter of aspartame-sweetened diet drink releases fifty-six milligrams of methanol. What does this mean in terms of toxicity? Not much, as we can see when we compare the amount to other sources of methanol in our diet. Methanol occurs naturally in fruit juices, averaging about 140 milligrams methanol per liter, and wine can have up to 320 milligrams per litre. Aspartame's opponents maintain that the body handles methanol differently when it is ingested along with other alcohols, such as ethanol, which are found in juices or wine.

Their argument goes as follows. Methanol itself is not much of a problem, but it is metabolized by enzymes in the body to become highly toxic formic acid. This is true. The same enzymes metabolize ethanol. This is also true. The enzymes prefer to work on ethanol, so when both ethanol and methanol are present in the blood, the enzymes will be busy with ethanol and will leave methanol alone. The methanol is then excreted before it can do any damage. Still true. But when there is no ethanol around, the enzymes have a chance to work on methanol and convert it to formic acid. True again. And it is this formic acid that causes "methanol toxicity." This is where the argument becomes murky. Where is the evidence that formic acid levels in the blood rise with aspartame ingestion? A search through the medical literature reveals no such finding. But several studies present data indicating that formic acid levels do not change even after consumption of large doses of aspartame. The same is true for blood levels of methanol.

Actually, aspartame just may well be the most widely researched food additive ever to have come on the market. As with any other newly introduced substance, reports of adverse reactions were expected since no amount of testing can preclude idiosyncratic reactions in a small minority of the population. In reality, the number of such reports has been small. Over seventy million people in North America use aspartame on a regular basis, yet the number of reported complaints average only around 300 per year. The majority of complaints (67 percent) refer to headaches, dizziness, visual difficulties, and mood alterations. Gastrointestinal problems (24 percent) and allergic symptoms such as hives, rashes, and swelling of tissues (15 percent) have also been reported. On occasion seizures have been linked with aspartame exposure. In most instances, these difficulties were noted only when aspartame intake far exceeded normal use.

Double-blind challenges have been carried out with aspartame. At Duke University, in one of the best-designed such studies, the effects of a single large dose of aspartame in people who had claimed to be sensitive to the substance was investigated. The results showed no difference in headache frequency, blood pressure, or blood histamine concentrations (a measure of allergenic potential) between the experimental and control groups.

In another study at the University of Illinois involving diabetics, subjects in the placebo group actually had more reactions than those in the aspartame group. On the other hand, surveys by physicians in headache clinics reveal that aspartame precipitates headaches in about 8 percent of the patients they see. This kind of conflicting data is characteristic of the research on the possible side effects of aspartame. Reported anecdotal experiences are not confirmed by carefully controlled scientific studies. This of course does not mean that the problems are not real, but it does imply that in many cases the symptoms may not be caused by aspartame. People get headaches, upset stomachs, and aches and pains of all kinds on a regular basis for no easily determined reason. If they recall having consumed aspartame when one of these ailments strikes, the sweetener may be judged to be guilty by association. This is even more likely if people are familiar with some of the adverse publicity that aspartame has received.

Perhaps the best double-blind study ever carried out in this area failed to find any aspartame effect. Dr. Paul Spiers and colleagues at the Massachusetts Institute of Technology gave subjects aspartame at a dose equivalent to more than a dozen diet drinks a day, and found no difference in brainwaves, mood, memory, behavior, or physiology. Reports of headaches, fatigue, and nausea occurred with equal frequency in the aspartame and placebo groups. Opponents cast a shadow on this study because of financial support from the industry. But where should researchers seek funds for sweetener research? From a light bulb manufacturer? Receiving a grant does not necessarily mean that a researcher has been bought off.

Certainly there have been some studies that have found adverse reactions to aspartame. At least one study has confirmed allergic symptoms such as hives and swelling in sensitive individuals. It is unclear how the allergy comes about, since none of the components of aspartame are believed to be capable of producing allergic reactions. It has been suggested that diketopiperazine, a compound that forms when aspartame decomposes, may be responsible. Some consumers, then, really may have some adverse reaction to aspartame. There does seem to be a little too much smoke for there to be no fire at all.

But if you listen to the likes of Betty Martini, we're not talking about fighting a small fire, we're battling a blazing inferno. Excuse me, it should be "Dr." Martini, as she now signs the voluminous literature with which she floods the Web. After all, Martini does have an honorary doctorate in humanities from some unaccredited religious institution. She has no degree in science but is convinced that "aspartame disease" is ruining millions of lives around the world. Martini spews out a few scientifically legitimate facts and then proceeds to mangle them beyond recognition. She is absolutely passionate and wholeheartedly believes in the cause.

For example, Ms. Martini correctly states that methanol in the body is metabolized to formaldehyde, and then comes up with the astounding statement that "methanol/formaldehyde is the strongest organic base in the living organism and is a polymerizing agent, turning tissue into plastic." This is absurd beyond words. She also readily concludes, without any evidence, that Baltimore Oriole pitcher Steve Bechler's death, which was blamed on using ephedra for weight control, was in reality caused by aspartame, which "destroys the heart." Oh, and the problems attributed to silicone breast implants were of course also due to aspartame. According to Martini, this fact never came to light because potential informants were paid off.

Martini doesn't realize it, but such silly rants coupled with her venomous inflammatory rhetoric actually serve to undermine her cause. When confronted by her "facts," most of which amount to overwhelming drivel, the tendency may be to throw the baby out with the bath water. Although Martini is in desperate need of some repairs to her mental machinery, she does deserve credit for having gathered a massive amount of anecdotal evidence about supposed adverse reactions to aspartame. Most of her claims, such as "Athletes everywhere appear to be dropping like flies because aspartame damages cardiac conduction," can be dismissed as juvenile. But there are some that merit attention. While the majority of scientific studies have given aspartame a clean bill of health, a few have suggested that aspartame may indeed be linked to headaches, visual problems, and mood disturbances. It is worth noting that while all of the industry-sponsored studies have found aspartame to be safe, 90 percent of independently sponsored studies have found some potential for adverse effects. The biggest concern, of course, is when the specter of potential carcinogenicity is raised.

Dr. John Olney of Washington University, whom we have already encountered because of his anti-MSG crusade, was the first to raise the alarm by claiming that an increase in brain tumors among the American population paralleled the introduction of aspartame. Others demonstrate that the increase began about eight years before aspartame was introduced and has levelled off while aspartame use has skyrocketed. Betty Martini's answer to this divergence of opinion is that the FDA and many researchers have been bought off by the manufacturers of the sweetener. She would probably say the same thing about researchers in California who published a 1997 paper in the *Journal of the National Cancer Institute* describing how they collected information on aspartame exposure from fifty-six brain cancer patients, all under age nineteen, and compared their intake with that of ninety-four controls. Patients with tumors were no more likely to have consumed aspartame, and maternal consumption did not elevate risk in breastfed or in nonbreastfed children.

More recently, Dr. Morando Soffritti, a respected cancer researcher at the European Ramazzini Foundation of Oncology and Environmental Sciences, stirred the cancer pot with his widely publicized study involving 1,900 rats fed varying amounts of aspartame over their lifetimes. He found that the equivalent of three liters of diet soda a day was associated with an unusually high rate of lymphomas and leukemias in the animals. Scary stuff. But there were some curious inconsistencies in the data. Female rats, for example, showed no dose-response relationship in terms of cancer, and some of the longest-lived rats were the ones that consumed the most aspartame, equivalent to 1,750 cans of diet soda a day. Nevertheless, regulatory agencies such as the European Food Safety Authority decided to examine Soffritti's evidence carefully to see if some change in regulations was warranted. After reviewing the data, the independent panel of scientists found that the study had wrongly concluded that aspartame led to higher rates of leukemia and lymphoma in the rats, and they highlighted a number of methodological flaws. Indeed, the panel of toxicologists asserted that it saw no reason to undertake any further review of the safety of aspartame.

Just as this news hit the press, another study by the US National Instititutes of Health was released and sent the Martini clan scurrying for a thesaurus to find appropriate denigrating words. The study was huge, involving over half a million men and women who, in the mid-1990s, filled out detailed food questionnaires that allowed researchers to calculate the amount of aspartame in their diets. Many of these people in later years developed some sort of cancer, but no connection was found between aspartame consumption and type or number of tumors.

Currently, there is no human epidemiological evidence linking aspartame to cancer. If such a link existed, it would have shown up in a huge study published in 2007 in the *Annals of Oncology*. Italian researchers led by Dr. Silvano Gallus collected data over thirteen years from across the country, comparing sweetener consumption in cancer patients versus healthy controls. There was absolutely no correlation found between aspartame consumption (or that of any other sweetener) and nine commonly occurring cancers. We can also note that diabetics number among the largest consumers of aspartame, and no link between the sweetener and any form of cancer has ever been found in this group. Without a doubt, there are people who may suffer adverse reactions with high consumption, but for the vast majority, aspartame does not pose a serious risk to health.

This view is backed up by perhaps the most comprehensive review of aspartame ever conducted, published in the September 2007 issue of *Critical Reviews in Toxicology*, a peer-reviewed journal. A panel of eight scientists with expertise in toxicology scrutinized more than 500 studies and reports about aspartame, including the original Ramazzini study, as well as a follow-up in which Soffritti and colleagues suggest that doses of aspartame comparable to those consumed by human beings cause cancer in rats. The panel found a number of methodological flaws in the Ramazzini trials and concluded that "the weight of existing evidence is that aspartame is safe at current levels of consumption." These levels, roughly five milligrams per day per kilogram of body weight, are way below the government-approved upper intake level of fifty milligrams per day per kilogram of body weight.

Does this extensive review of aspartame close the door on the controversy? Not a chance. The anti-aspartame crowd was quick to point out that the panel's work was funded by Ajinomoto, a company that makes aspartame. These people implied that the panel's conclusions could not be trusted. Not so. Members of the panel were unaware of the origin of the funding until the final manuscript had been submitted, and Ajinomoto was unaware of the panelists' identities. And after all, who would be interested in funding a review of aspartame? A car manufacturer? No. Those who potentially have something to gain from such a review. Still, the issue of who funded the research does not invalidate the process or the findings.

Sucralose ("Splenda") is a more recent addition to the sweetener market and has shown itself to be a worthy rival of aspartame in sales—and in controversy. Again, its discovery at Queen Elizabeth College, University of London in 1976 was accidental. Professor Leslie Hough was involved in some collaborative research with Tate & Lyle, a British sugar company interested in finding novel uses for its product. Hough had asked graduate student Shashikant Phadnis to make some chlorinated sugars and to test them. The foreign student apparently confused "test" with "taste" and discovered the incredible sweetness of his new molecule. Sucralose, as the new compound came to be called, turned out to be 600 to 1,000 times sweeter than sugar, depending on what it was added to.

Sucralose is highly water soluble, as well as stable to heat and acid, making it easy to use in diet drinks and baked goods. Sucralose is so sweet that much less of it than sugar is needed for comparable sweetness. Sugar, however, provides not only sweetness but also bulk in bakery products. The problem of missing bulk is solved when sucralose is combined with maltodextrin, a bulky starch; this mixture can be substituted for sugar, measure for measure. Sugar, however, also gives a brown color to baked goods, so some sucralose-sweetened goodies may look rather anemic.

As one might expect, safety testing of sucralose has been extensive. For fifteen years it was subjected to a battery of short-term and long-term animal feeding studies. The results were conclusive. About 85 percent of a sucralose dose was excreted unchanged, and even the small percentage that was metabolized yielded compounds that were also excreted. All of the chlorine atoms in the sucralose that animals were fed could be accounted for in their excreta. Any concerns about storage in the body or interference with metabolic pathways are unfounded. As an added benefit, unlike sugar, this sweetener has no detrimental effect on the teeth. While our bodies cannot break down sucralose, microorganisms in water and the soil readily do so. In other words, sucralose is biodegradable and poses no environmental hazard. As with any substance, there can be no absolute guarantee about the safety of sucralose for everyone. Any food or additive, be it peanuts or apples, aspartame or sucralose, can cause a problem for some people. But reactions to sucralose are rare indeed.

Another of sucralose's attributes is that it leaves no bitter aftertaste, but unfortunately the same cannot be said about some of its advertising hype. "Made from sugar, so it tastes like sugar," is the popular slogan meant to convey the message that sucralose is somehow more "natural" than other artificial sweeteners. Natural does not equate with safe, but that is not the issue. What a substance is made from is irrelevant; what matters is what the final product is. Its properties are determined not by its ancestry but by its molecular structure. Hydrogen gas, for example, can be made from water but it would be absurd to suggest that it therefore has the same safety profile. It's a different substance, just like sucralose is different from sugar. Incorporation of three chlorine atoms into the sugar molecule converts it into a totally new substance. We know sucralose is safe because it has been extensively tested, not because it is made from sugar.

There is one last comment to be made about artificial sweeteners in general. Their sales have increased dramatically in the last couple of decades, but so has obesity. Artificial sweeteners may be of great help to diabetics but they are definitely *not* the answer to our weight-control problems.

IMPROVING TASTE WITH ARTIFICIAL FLAVORS

One strawberry ice cream brand boasts that it is made with "all natural flavors," while another, cheaper version lists "artificial flavor" among its ingredients. Which one would an ice cream lover expect to deliver a more authentic strawberry experience? Surely, the natural flavor must be superior! How can "artificial" be expected to match the real thing? Well, surprise, surprise! The artificial strawberry flavor may actually be closer to the taste of real strawberries than the "natural" flavor. And therein lies some interesting chemistry.

Of course, only a real strawberry tastes like a real strawberry. That's one of the reasons why consumers seek out "natural flavor" on the label. But are they really getting natural strawberry flavor? Not likely! And how do we know that? Well, the amount of natural strawberry flavoring sold annually around the world exceeds the amount that could possibly be produced from all the strawberries grown in the world by a factor of about three. So what sort of magic is going on here? Here's the surprise. The term "natural flavoring" on the label of the strawberry ice cream does not mean that all the components have to come from strawberries. It does, however, mean that all the components have to come from natural sources. The ideal of course would be to use real strawberry juice, but it is too expensive, and there just isn't enough of it to meet the world's craving for strawberry flavor.

This is where a flavor chemist, or "flavorist," enters the picture. His or her task is to reproduce the flavor and aroma of strawberries (aroma is a major contributor to taste) by blending together readily available natural components. Real strawberry juice is used as the base, but other substances such as the essence of cloves or extract of orrisroot are added with the hope of mimicking an overall strawberry aroma and taste. The result may be a close approximation of the desired flavor, but it will not be identical to it. It cannot possibly be. More than 300 compounds have been identified as components of natural strawberry flavor, and this particular mix cannot be reproduced exactly by blending natural substances that do not originate from strawberries. But blending artificial or "synthetic" substances can come close.

The creation of an artificial strawberry flavor requires the expertise of analytical chemists, synthetic organic chemists, and flavorists. First, an analytical chemist identifies the compounds that make up the flavor of strawberries. This is no easy task, given the large number involved. The advent of modern instrumentation—specifically gas chromatography, mass spectrometry, and nuclear magnetic resonance (NMR) spectrometry—has, however, greatly facilitated the job. A gas chromatograph can separate the components of a mixture, the mass spectrometer can then determine the molecular weight of each component as well as offer important clues about its composition, and an NMR spectrometer can pin down the specific molecular structure. Once a compound has been identified, an organic chemist can attempt to synthesize it from simple raw materials that may be derived either from petroleum or from plant or animal sources. If the synthesis is successful, the compound cannot in any way be distinguished from the one that is made by the strawberry plant.

For example, methyl butanoate is one of the compounds that dominates natural strawberry flavor. It can be readily made in the lab from methanol and butanoic acid, but according to regulations, it then has to be termed "synthetic" or "artificial," even though it is identical in every way to the methyl butanoate that can be extracted from strawberries. In theory, each of the more than 300 compounds in strawberry flavoring could be synthesized and combined in appropriate amounts to reproduce natural strawberry flavor. It would have the same taste and have the same safety profile as real strawberry extract but would still have to be called "artificial." Synthesizing all the compounds involved in the strawberry flavor would be a monumental job. It would also be an unnecessary one, since relatively few of these make a major contribution to the overall aroma and taste.

Why then not select the ones that are really important and create an artificial flavor from these? Enter the flavorist. From the hundreds of bottles of pure chemicals on his shelves synthesized by organic chemists, he selects roughly the ten to twenty that, according to analytical chemists, have been found in strawberries and are most likely to contribute to flavor. And now art begins to blend with science. The flavorist sniffs, tastes, mixes, adds, subtracts, or substitutes compounds until a quality strawberry flavor is achieved. This "artificial" strawberry flavor is in fact composed of compounds that are actually found in strawberries. It will still not have the same flavor as fresh strawberries, because that flavor is due to a symphony of numerous compounds, many of which make subtle contributions.

If the flavorist is not completely satisfied with his creation, he may try improving it by adding flavor compounds that are not actually found in strawberries. These may come either from the more more than 6,000 compounds that have been identified as contributors to the flavor of various foods, or from an array of synthetic compounds chemists have formulated to produce flavors not found in nature. For example, 3-methyl-2-butyl ethanoate is not found in nature but has a decidedly fruity taste. It is used in flavoring Juicy Fruit gum, but can be used to improve other flavors as well.

While it is impossible to guarantee the absence of any adverse reaction to some component in an artificial flavor (or in a natural flavor), it is reassuring that the compounds available to flavorists have undergone review by various health authorities and have been deemed to be "generally recognized as safe" (GRAS). But there is a final point to remember: artificial flavorings are mostly used in processed foods, which should not be the mainstay of a nutritious diet. Your taste buds may react the same way to naturally occurring gamma-undecalactone in fresh peach juice as they do to its synthetic analogue in a peach-flavored drink, but the juice is certainly a better nutritional choice.

PREVENTING BOTULISM WITH NITRITES

Every Fourth of July a crowd of more than 30,000 gathers in front of Nathan's Famous Hot Dog stand on Coney Island in New York to enthusiastically cheer on the competitors in the world's most celebrated eating contest. In 2006, spectators watched in awe as reigning champion Takeru Kobayashi polished off fifty-three and three-quarters franks and buns in twelve minutes to break his own world record. Kobayashi withstood a late surge by American Joey Chestnut, who devoured a spectacular fifty-two hot dogs in the allotted time. A dozen others managed to stuff themselves with more than twenty wienies. It goes without saying that such epic achievements require extensive training that involves wolfing down hot dogs year round. These feats of gustatory extreme should not go unrecognized by the scientific community: these competitive eaters constitute a unique experimental group. Their nitrite intake exceeds that of the average person by orders of magnitude. Given the controversial nature of this food additive, the health status of the eating champs merits monitoring.

One of the most deadly substances known to mankind is produced by a species of bacteria known as *Clostridium botulinum*. Seven million times more toxic than cobra venom, botulin poisons its victims by blocking the action of acetylcholine, a neurotransmitter. Symptoms can range from double vision and difficulty in swallowing to paralysis and death. The spores of this organism lurk in many foods and under the right conditions (lack of oxygen and low acidity), become active and liberate their toxin. Sausages are the classic example of the type of food that can be affected, and the word *botulism* derives from the Latin *botulus*, meaning "sausage."

Luckily, botulism can be prevented by the appropriate use of sodium nitrite, a discovery that came about in an accidental fashion. Salting is an ancient method of preservation based upon the ability of sodium chloride to kill bacteria by drawing out much of their water content. About five hundred years ago, some clever cook noted that the effectiveness of salt in preserving meat depended on its source. Furthermore, salt that worked particularly well also improved the meat's flavor and color. The secret turned out to be an impurity, potassium nitrate, more familiar to us as saltpeter (a major ingredient used to make gunpowder). Actually, the real secret isn't nit*rate,* it is nit*rite.*

Some bacteria in meat are resistant to salt and have the ability to convert nitrates into nitrites. Food processors soon realized that it was more efficient to use sodium nitrite directly as a preservative. The situation is even more complicated, because the true active ingredient isn't nitrite, but its decomposition product, nitric oxide. It is this substance that produces the characteristic reddish pink color of cured meats and improves their flavor. Nitric oxide reacts with myoglobin, one of the compounds responsible for the color of meat, to form pinkish nitrosomyoglobin. It also prevents the iron atom incorporated in myoglobin from catalyzing the oxidation of fats, which causes meats to turn rancid. But nitric oxide's most important function is to destroy botulin-producing bacteria. There is concern, though, that bacteria may not be all that nitrites destroy.

Worry about the relatively small amounts of nitrites used as food additives began in the 1960s when researchers noted that domestic animals fed fish meal preserved with nitrites were dying of liver failure. The problem was traced to a group of compounds called nitrosamines, which formed by a chemical reaction between the naturally occurring amines in the fish and sodium nitrite. Nitrosamines are potent cancer-causing agents and their possible presence in human foods became an immediate concern. An examination of a wide variety of foods treated with nitrites revealed that nitrosamines could indeed form under certain conditions. Fried bacon, especially when "done to a crisp," consistently showed the presence of these compounds. So did hot dogs. And to everyone's great horror, beer was also contaminated! What was going on?

In the case of meat, there was no great mystery. When proteins break down during cooking they can yield amines, which in turn react with nitrites to form nitrosamines. But the beer issue took some investigating before the problem was traced to the flame-drying of malt. When air is heated, its nitrogen and oxygen components can react to form various nitrogen oxides, which in turn can react with amines in the malt to produce nitrosamines. Much to their relief, brewers were able to alter the process and prevent direct contact of malt with flame.

Curbing nitrosamine formation in cured meats turned out to be more of a challenge, but producers started to add ascorbic acid (vitamin C) or its close relative, erythorbic acid, because these compounds interfere with the reaction of nitric oxide with amines. They also enhance the preservative effects of nitrite by promoting its conversion to nitric oxide, allowing for less nitrite to be used. In recent years the nitrite levels in foods have been continuously reduced to the point where most products now contain less than 100 parts per million.

The possible formation of nitrosamines within the human body is another issue. We ingest both nitrites and amines in our food, and the possibility exists that these can combine to form carcinogenic substances under the acidic conditions found in the stomach. Some epidemiological studies suggest that this process is more than a theoretical possibility. A recent Swedish investigation found that eating more than three servings of processed meats a week was associated with a significant increase in stomach cancer when compared with consuming less than 1.5 servings a week. Unprocessed red meats such as hamburger, poultry, or fish did not show such a relationship.

There is still another facet to the nitrite issue. Most of the nitrite in our body does not come from processed foods. Nitrates occur extensively in vegetables such as lettuce, spinach, beets, and celery and are converted to nitrites by enzymes in our body. Indeed, nitrite added to food represents only about 10 percent of our total nitrite intake. But before we get all worked up about nitrates in our salad, let's remember that numerous epidemiological studies have shown that eating vegetables reduces the risk of cancer. Whatever risk naturally occurring nitrates may represent is greatly outweighed by the various cancer-inhibiting compounds found in vegetables.

Nitrites used as additives in processed foods may be a greater concern than nitrites produced in the body because they arrive in the stomach in a more concentrated fashion. However, to really nail down the effects of nitrite additives on health, we need to study high-nitrite consumers. So let's watch the hot dog–eating contestants closely for a few years. Another possible benefit might come from watching these contestants rip hot dogs apart, stuff them into their mouths, and wash them down with a chaser of buns soaked in water. The bizarre sight just may reduce your risk of nitrosamines by taking away your appetite for hot dogs.

PRESERVING WITH SULPHITES
AND PROPIONATES

Sulphites prevent wine from spoiling. They allow us to munch on grapes in the middle of winter. They make our pizza crust crispy. They preserve our dried fruits. But they can also create misery for asthmatics, and on rare occasions, they can even kill. A class of chemicals widely used in food and beverage processing, sulphites have the ability to release sulphur dioxide, a multifunctional reagent.

Our first encounter with sulphur dioxide takes us back to Homer's writings in the eighth century BC. The illustrious Greek poet refers to the burning of sulphur to fumigate homes to rid them of pests. Just what kind of pests he was talking about isn't clear, but the sulphur dioxide gas that forms when sulphur is burned is indeed very nasty. If you've ever experienced the choking odor of a burning match, you know what we are talking about. In a high enough concentration, the gas does more than irritate, it can be lethal to people, animals, plants, and microbes. Indeed, it was the need to control the proliferation of undesirable microbes in wine that led to the first use of sulphur dioxide as a processing aid by the ancient Romans. Of course, they had no idea about microbes, but we can guess at how sulphur dioxide treatment might have been introduced.

A classic sign of wine "going off" is a sour or acidic taste. The culprits are bacteria that produce lactic or acetic acids from the sugar, malic acid, or ethanol present in wine. These bacteria can hitch a ride aboard fruit flies that are attracted by the sweet taste. Some clever ancient vintner probably noted that the more flies that buzzed around the fermentation vessels, the greater the chance the wine would go off. Realizing that burning sulphur would eliminate such pests, he experimented with fumigating the vessels and was rewarded with higher-quality wines. And we have been happily treating wine with sulphur dioxide ever since. Not all wine is treated in this fashion, but this is of little comfort to those who experience negative reactions to sulphur dioxide because natural fermentation processes also produce this compound.

Fumigating wine containers with sulphur dioxide turned out to be useful in more ways than the ancient Romans ever imagined. Thanks to modern chemistry, we now realize that sulphur dioxide is capable of doing a lot more than controlling undesirable bacteria in wine. It can react with dissolved oxygen to form sulphates. This is very useful because some of the vinegar-forming bacteria are resistant to sulphur dioxide, but they need oxygen to convert ethanol to acetic acid. No oxygen, no acetic acid. Even in the absence of bacteria, oxygen is a problem. It can oxidize ethanol into acetaldehyde, which can react with more oxygen to produce acetic acid. Not only does sulphur dioxide mop up oxygen, it also reacts with acetaldehyde to form an adduct that does not go on to form acetic acid. So, the "off" taste that acetaldehyde contributes to wine is also reduced.

Although yeast is needed to ferment sugar into alcohol, there are some rogue yeasts that can also contaminate wine and produce off flavors. These too can be controlled with sulphur dioxide. How do they get into wine? In one analysis a ton of wine grapes was found to contain seven pounds of dirt, one mouse nest, 147 bees, 98 wasps, 1,014 earwigs, 1,833 ants, 10,899 leafhoppers, and an assortment of bird droppings. It's easy to see why the disinfectant properties of sulphur dioxide come in handy.

We're still not done. Grape juice contains phenols, and these react with enzymes released when grapes are crushed, yielding brown pigments that discolor wine. Sulphur dioxide can inactivate these enzymes. Like grapes, numerous other fruits and vegetables are also

susceptible to such "enzymatic browning," with cut apples, potatoes, and lettuce being classic examples. That's why restaurants used
to spray salad bars with a dilute sulphite solution. This practice was
stopped when it became evident that some people, almost all of them
asthmatics, experienced allergic-type reactions when exposed to
sulphites. One unfortunate woman almost died after eating french
fries soaked in a sulphite solution. Aware of her sulphite sensitivity,
she avoided wine and foods that listed sulphites on their labels, but
never thought of restaurant fries as a possible source. Luckily, quick
treatment with adrenaline in a hospital saved her life. A young girl
who had eaten guacamole from a salad bar that had repeatedly been
treated with sulphites to "freshen" it, wasn't so fortunate. Paramed-
ics were unable to revive her after she collapsed. Estimates are that
about 5 percent of asthmatics, or roughly a million people in North
America, show some sulphite sensitivity and that sulphites have been
responsible for about twenty fatalities.

Although sulphite use on salad bars is now *verboten*, sulphites still
appear in many foods. They meet so many food-processing needs
that they are very difficult to replace. Besides preserving dried fruits,
keeping grapes from rotting during storage, preventing shrimp from
forming "black spot," and fighting browning in fruits and vegetables,
they can also improve the baking properties of flour by altering the
structure of gluten, the protein found in wheat. Sulphites are obviously useful chemicals, but they are also the only approved food
additives known to have directly caused death. Of course, naturally
occurring compounds can also be accused of the same crime. Just
as people who have potentially life-threatening sensitivities to peanuts, shellfish, sesame seeds, and numerous other substances have
to learn to avoid these foods, sulphite-sensitive individuals have to
learn about all the places where sulphites can be lurking, such as in
maraschino cherries, sauerkraut, crackers, french fries, and—*sacre
bleu!*—in wine.

Propionates are found in a variety of breads and bakery products,
where they prevent molding. Not only can molds make for unsightly

green splotches, some produce decidedly dangerous compounds. That's why we add preservatives such as calcium propionate to bread. It prevents the growth of molds while allowing yeast to flourish. And that's not all that calcium propionate does. It also inhibits the formation of "rope" in bread. The spores of certain bacteria, such as *Bacillus mesentericus*, are often present in flour and germinate under the moist, warm conditions needed to make bread rise. These bacteria are not harmful to humans but they change the texture of the dough and produce sticky, yellow stringy patches that make for an unpalatable bread. Propionates prevent this from happening.

Are propionates safe to eat? Sure they are. Food producers can't just randomly add chemicals to their products. Additives are strictly regulated and must have a clearly demonstrated benefit with minimal risk before they are allowed into general use. In the case of propionates, safety is not hard to demonstrate. These compounds cruise through our bodies all the time, and they don't have to be introduced through bread. Bacteria in our intestine feed on fiber, the indigestible part of fruits, vegetables, and grains, and convert it into a variety of compounds that include propionic acid. This is then absorbed into the bloodstream. Far from being harmful, some studies have shown that such short-chain fatty acids can reduce the risk of colon cancer and may even be helpful in other diseases of the digestive tract.

Propionates, as derivatives of propionic acid are called, also occur naturally in our food supply. Perhaps the best example is Swiss cheese. The texture and flavor of this cheese is due to the addition of a starter culture that includes the bacterial species known as *Propionibacter shermanii*. These bacteria break down some of the fat to produce carbon dioxide gas, which explains the presence of holes in the cheese. They also produce propionic acid, which is responsible for some of the characteristic nutty flavor. Swiss cheese contains roughly 1 percent propionates by weight, far more than the amount used as a preservative in bread. So with all these propionates everywhere, it is little surprise that they end up in our blood, and even in our sweat. But they present no risk, except to molds.

PRESERVING WITH VIRUSES

Viruses, by and large, are nasty little organisms. Biologically, they are relatively simple, consisting of nothing more than bits of genetic material wrapped in a protein coat. They can reproduce, which is one of the hallmarks of living organisms, but there is debate about whether or not viruses can be classified as living creatures. Why? Because they cannot reproduce on their own. Viruses, however, are very adept at invading living cells, inserting their genetic material into the cell's reproductive machinery, and forcing it to crank out more viruses. As the viral load builds up, the host cells are altered or killed, and illness ensues. What sort of illness? Well, that depends on the virus. Some viruses may do no more than annoy, by causing benign conditions such as warts, the common cold, or chicken pox. But others can devastate health by triggering smallpox, rabies, SARS, cervical cancer, and AIDS.

Should anyone then be surprised that eyebrows are raised when the US Food and Drug Administration approves the spraying of a viral cocktail on certain meat products, such as hot dogs and sliced cold cuts? Actually, to say that this practice has raised eyebrows is an understatement. Some activists who already regard the FDA as a public enemy went into frenzied mental gyrations at the prospect of adding viruses to our food supply. Genetic modification is bad enough, they proclaim, but now a callous FDA has thrown all caution to the wind regarding food safety. Are they right?

There are viruses, and then there are viruses. Some can infect human cells, others can only attack bacteria. And this is the crux of the matter. The "virus cocktail" approved by the FDA only invades bacteria, and more specifically, only *Listeria monocytogenes*. And this is one nasty bacterium. Named after Joseph Lister, the British surgeon who first recognized the need to keep microbes out of the operating room, *Listeria* can turn up in our food supply and cause a great deal of human misery. The bacteria lurk in the soil and in vegetation, so it is easy to see how they can find their way into animals and humans. Unpasteurized milk, soft-ripened cheeses, raw and smoked fish, uncooked hot dogs, cold cuts, and raw vegetables can all harbor *Listeria*.

Once they infect the body, these microbes can be pretty wicked. Lucky people may get away with no more than flu-like symptoms, fever, headache, vomiting, cramps, and a touch of diarrhea. But if you are unlucky, the bacteria can find their way into the bloodstream and cause blood poisoning (septicemia) or, if they invade the brain or spinal cord, meningitis. These conditions can be lethal if not appropriately treated with antibiotics. The young, the elderly, the immune-compromised, and pregnant women are most susceptible to infection. Listeriosis in early pregnancy can cause miscarriage or stillbirth, which is why pregnant women are often advised to stay away from foods such as sliced, packaged meats. To complicate matters further, victims may not easily connect their illness to food because the symptoms of listeriosis can present anywhere from a couple of days to three months after consuming a contaminated product. And to top off the concerns, *Listeria* is one of the few bacterial species that can multiply at refrigerator temperatures.

Obviously, controlling *Listeria* in our food supply is of great importance. Washing vegetables well, cooking meat products thoroughly, and avoiding unpasteurized milk (forget the unsubstantiated health claims about the benefits of raw milk) can go a long way, but cannot eliminate all risk. After all, we don't wash our sliced turkey or chicken after opening the package. And this is where viruses can help. Specifically, viruses called bacteriophages, the name coined in 1917 by their discoverer, Félix d'Herelle, from the Greek *phagein*, meaning "to eat." Although Montreal-born d'Herelle gets credit for the discovery, he was not the first to observe these bacteria-eating organisms. As early as 1896, British physician E. Hanbury Hankin passed water from the Ganges River through a fine porcelain filter and noted that the filtrate had antibacterial properties. About twenty years later, bacteriologist Frederick Twort actually managed to isolate microscopic entities that destroyed bacterial cultures but never pursued the work.

Remarkably, Félix d'Herelle had no formal education but managed to set up a laboratory in his home and train himself as a microbiologist; he pursued numerous research interests, including the development of yeast strains to make cheap whisky from rotting fruit. His major discovery came during the time he worked as a volunteer at the Pasteur Institute in Paris, where he was asked to investigate an epidemic of dysentery that was raging in a cavalry squadron.

D'Herelle didn't know exactly how the disease was spread, but suspected that it was through fecal matter. He took samples from the soldiers, placed these in a filter with microscopic pores and passed water through it. The idea was to see if the filtrate had any sort of infectious agent. Much to d'Herelle's surprise, not only was the liquid free of any such substance, but when it was mixed into a bacterial culture it caused the formation of clear spots, indicating that bacteria had been destroyed. "In a flash I had understood what caused my clear spots," d'Herelle later recalled. "It was in fact an invisible microbe . . . a virus parasitic on bacteria."

D'Herelle managed to isolate these parasitic viruses and suggested they could be used to treat bacterial infections in humans and animals. Some early successes inspired George Eliava, a colleague of d'Herelle's at the Pasteur Institute, to return to his native Georgia in Eastern Europe, and with d'Herelle's help, set up a "bacteriophage" institute. Indeed, the Eliava Institute has become a world leader in phage therapy, producing a great deal of research that has largely been ignored in the West. But now bacteriophage preparations may help us control the spread of listeriosis. Human cells do not have receptors for these viruses, so we cannot be infected. In fact, we are exposed to them all the time; bacteriophages are found wherever bacteria are found, in the soil, in our water, in our food. There are some concerns that viral proteins may cause allergies or that the viruses may affect some of the beneficial bacteria in our gut, but those concerns are theoretical. Here is what is factual: more than 500 people a year in North America die from listeriosis. Bacteriophage treatment can help reduce this toll. So we see not all viruses are nasty.

PRESERVING WITH RADIATION

I t was a publicity stunt, to be sure, but nevertheless it was thought provoking. Reporters and photographers stood by as David Corbin, chairman of the Texas-based Sadex Corporation, sat down to a meal of spinach. Not ordinary spinach, mind you. These leaves had been inoculated with millions of colonies of the dreaded *E. coli* 0157:H7 bacteria, the nasty bugs that had terrorized the spinach-eating world in that famous 2006 episode. Corbin, however, was not worried. His spinach had undergone electronic pasteurization, meaning that it had been subjected to irradiation with a beam of microbe-destroying electrons. Corbin experienced no ill effects and made his point. If ready-to-eat foods like spinach were irradiated before delivery to stores, the risk of bacterial food poisoning could be dramatically reduced. Needless to say, the Sadex Corporation is in the food irradiation business.

Foreheads always start to wrinkle at the very mention of any process that involves "radiation." Images of Hiroshima, Chernobyl, and the yellow "Danger: Radiation Risk" symbol immediately flash through minds. For some, the thought of eating irradiated foods may even raise fears of becoming radioactive and glowing in the dark. But such fears are irrational, a consequence of a lack of understanding of what is meant by radiation.

If we choose the simplest possible definition, radiation is the propagation of energy through space. Light coming from a light bulb is radiation. So is heat from a radiator. When we bask in the sunshine, get an X-ray, or receive Cobalt-60 treatment for cancer, we are exposed to radiation. In the latter case, the radiation is in the form of gamma rays generated by the spontaneous decomposition of atoms of Cobalt-60. This is what is known as *radioactivity*. Clearly, the risks, or indeed the benefits, of radiation cannot be evaluated without proper context. It is the type of radiation and extent of exposure that determine risk.

Visible light, unlike X-rays or gamma rays, does not have enough energy to break chemical bonds, so you don't have to worry about your bedside lamp inflicting damage on your molecular constituents. X-rays, on the other hand, can cause significant changes in molecules. Even here, though, dosage matters. While the risk associated with a single X-ray may be minimal, frequent CAT scans can disrupt normal cellular activity. In this case, such disruption is undesirable, but when it comes to irradiating foods, it is the very effect we are after. The goal is to throw a wrench into the cellular machinery of microbes and cause their demise.

As early as 1905, patents were issued for devices that used X-rays to kill bacteria in food. Later, gamma emitters and electron beam generators were developed, and by 1958 a number of countries were using such "irradiation" techniques to preserve various foods. Of course, there was immediate opposition to the technology, as there invariably is to any new venture. Irradiation was concocted by the nuclear power industry as a way of disposing of nuclear waste, some activists maintained. Others claimed it destroyed nutrients in food, gave rise to novel toxins, and provided an easy way for producers to cover up sanitation problems.

Let's get one thing straight. Eating irradiated food does not expose consumers to radiation. The insects and microbes that contaminate our food supply can testify to the lethal effects of radiation exposure, but the food does not become radioactive. Furthermore, Cobalt-60 is not a waste by-product of the nuclear industry, and X-ray machines and electron beam devices do not use radioactive materials. True, transporting radioactive cobalt and its disposal present some challenges, but appropriate technologies are available. Nutrient destruction in food is really a nonissue. Any treatment, whether it be cooking, canning, drying, or freezing will lead to some nutrient loss; irradiation, if anything, has less of an impact than these processes.

What about the novel "toxins," or "unique radiolytic products" that supposedly result from irradiating food? There is no doubt that exposing food to radiation results in some chemical changes— as does cooking. Most of the compounds formed upon radiation exposure are also found in cooked food, but not all. So far, 2-alkylcyclobutanones (2-ACBs) have been found only in irradiated foods, and concerns were raised in 2002 when European research- ers synthesized these compounds and tested them against cell lines in the laboratory. They found that 2-ACBs were capable of induc- ing mutations in cells, and that when fed to rats they promoted tumor formation in animals that had also been treated with a known car- cinogen. But the concentrations used were 1,000 times that found in irradiated foods, and the researchers themselves made clear that their data did not present any sort of indictment of irradiated foods. Benzene, a known carcinogen, can also result from radiation expo- sure but the amounts are inconsequential. Irradiated beef has about three parts per billion of benzene compared with the sixty parts per billion that occurs naturally in eggs. The fact is that over the last fifty or so years, numerous animal feeding studies of irradiated foods have been carried out, in many cases using extreme amounts. Dogs, rats, and mice have been fed irradiated chicken that made up 35 percent of their diet with no effect.

While the risks of irradiation are theoretical, the risks of food contamination are not. There are over eighty million cases of food poisoning a year in North America, with some 350,000 hospitaliza- tions and about 6,000 deaths. These could all be significantly re- duced with appropriate food irradiation. David Corbin's suggestion of protecting all spinach by irradiation is unnecessary, but perhaps his bravado will help change the negative image with which this technology has been unfairly burdened. What is needed is rational discussion, not the absurd and meaningless rantings of infomercial king Kevin Trudeau, who asserts that "irradiation changes the energetic frequency of food, giving the food a frequency that is no longer life-sustaining but, rather, toxic to the body."

COLORING WITH FOOD DYES

The young mother was really nervous. For the fourth time in twenty-four hours her infant daughter had filled her diaper with bright green poop! What new food had she introduced, her physician wanted to know? Nothing, was the reply, except for purple Kool-Aid, a beverage to which the little girl had taken a sudden fancy. Not a great nutritional choice, the doctor said, but purple Kool-Aid wouldn't produce green stools. Unconvinced, the mother began to search the Web for possible causes of "green poop." Luckily for her, a college student in Vermont had explored this very situation. He had become aware of the green spectacle after consuming copious amounts of Purplesaurus Rex Kool-Aid. Being a biochemistry major, he was sufficiently intrigued to carry out some basic research. Volunteers were not hard to come by. The word about the green emissions spread and e-mails detailing the results started to flood in. Sure enough, the purple Kool-Aid effect was real, and there was a dose-response relationship. One glass produced no result, twelve glasses yielded a decided greenish color, and one volunteer, who downed twenty-four glasses, for reasons known only to college students, reported a truly resplendent green production.

Purplesaurus Rex is colored with a mix of a blue and a red food dye. The blue coloring apparently reacts with yellow pigments in bile to produce a spectacular green that masks other colors. This was comforting information to our worried mother who now realized that the colorful effect she had noted was due to a harmless food dye.

Oh, I can see the eyebrows being raised out there. "What do you mean 'harmless' food dye?" Haven't these chemicals been accused of all sorts of crimes, ranging from causing hyperactivity in children to bronchospasm in asthmatics and cancer in rats? Yes, they have, but these accusations have to be carefully examined before we panic. Admittedly, food dyes have a checkered history with more than a few skeletons in the closet.

In the eighteenth and nineteenth centuries, unscrupulous merchants used various colored substances to spruce up foods that were spoiled or were of poor quality. Pickles were colored with copper sulphate, a process that caused misery to many and undoubtedly was responsible for some deaths. Toxic mercury and lead salts were used to dye candy, and thorn leaves were colored with copper acetate so that they could be sold as Chinese tea. Today, the situation is very different. Food additives have to go through a battery of safety tests before their use is allowed. Of course, regulations don't mean much to dishonest people. Witness the recent recall of a number of products in the United Kingdom that were tainted with Para Red or Sudan I, two dyes that cannot be legally used in foods because they have been shown to be carcinogenic in animals. These dyes were found in imported spices such as chili and cayenne peppers, which then showed up as ingredients in a variety of processed foods, including popular products like barbecue potato chips, chili sauce, and salmon pâté. Although the risk to humans is very small, nobody likes the idea of a carcinogen in the food supply.

Even with additives that have passed all the required safety tests, the possibility of an adverse reaction cannot be ruled out. Humans are biochemically individual, and the unexpected sometimes happens. A young physician was hospitalized on four separate occasions over a two-year period with abdominal cramps so severe that narcotics were required for relief. It turned out that he had a rare type of allergic gastroenteritis triggered by a food dye known as Sunset Yellow (FD&C yellow 6), found in foods such as Corn Bran cereal and Jell-O, which he regularly consumed. A somewhat more common reaction, although still rare, is to another yellow dye, tartrazine (FD&C yellow 5). Up to 20 percent of patients with asthma are sensitive to aspirin and suffer acute bronchospasms, swelling, and hives upon exposure. Roughly 10 percent of these people are also sensitive to tartrazine and have to check labels carefully for the presence of this dye.

In the 1970s Benjamin Feingold, a California pediatrician, raised the possibility that certain food additives, such as food dyes, can cause hyperactivity in children. His thesis was buttressed by testimonials from parents who claimed that junior went from devil to angel when an additive-free diet was introduced. Some suggested the results were due to wishful parental thinking or just a general improvement in diet caused by cutting down on processed foods. But now British researchers have found scientific evidence for the additive–hyperactivity connection. In a trial involving 153 three-year-olds and 144 eight- and nine-year-olds, half the children were given a mix of four food dyes as well as the preservative sodium benzoate dissolved in a fruit drink. The other half were given a placebo beverage that looked and tasted the same. The children, the experimenters, the parents, and the teachers who were asked to evaluate the kids' behavior were not informed about who was getting what. Final analysis of the results revealed a slight increase in hyperactive behavior, judged to be about 10 percent, in the group that consumed the additives.

What does this mean? Was one specific additive responsible? We can't tell. Was the effect due to some synergistic reaction between the additives, which would not be possible if they were consumed independently? We can't tell. Could it be that when added to food these chemicals show a different effect than when combined in a drink? We can't tell. But so what? Do we really need more evidence to get kids to cut down on soft drinks, candies, cakes, or sugar-laden gelatin desserts? I don't think so. Food industry spokespeople predictably claim that the study is methodologically flawed. "Natural food" advocates retort with an "I told you so," and continue to portray all additives as toxins. And governments pledge to scrutinize additives more vigorously and promise to introduce legislation to protect children from some of the more worrisome ones. But replacing brightly colored candies with anemic-looking sweets will not solve our nutritional problems. Urging children to eat apples, oranges, and nuts instead of processed, dyed foods will.

A more serious allegation is that certain food dyes, such as Red Dye #3 (erythrosine), may cause cancer. Very large amounts of this colorant have indeed caused thyroid tumors in male rats, but it is hard to estimate the relevance of this to humans. The industry claims that this dye, used in foods such as maraschino cherries, is hard to replace because, unlike other red dyes, it doesn't bleed out into its surroundings. There is hope, though, that a natural red dye, an anthocyanin isolated from radishes, will be up to the task. Indeed, many food processors are working on replacing synthetic dyes with natural ones because of the public's perception that these are safer. In this instance, that perception is by and large correct. Natural colors extracted from beet juice, annatto (a derivative of the achiote tree), grape skin, cabbage, turmeric, and paprika present no health issues. And, in what may come as a surprise to most people, the most widely used food colorant is "natural." What is it? Burned sugar! Caramel accounts for more than 90 percent by weight of all colors added to food and drink, with consumption of more than 200,000 tons a year worldwide. Your colas, soup mixes, chocolate cookies, and even some beers owe their color to caramel. It's safe, and of course you don't have to worry about any unusually colored bowel movement.

IMPROVING HEALTH WITH BACTERIA

In the late eighteen hundreds, the Battle Creek Sanitarium was unquestionably the place to be for people who needed to be cured of diseases they didn't have. It was there that Dr. John Harvey Kellogg and his staff catered to the rich hypochondriacs who, according to Kellogg, suffered from *autointoxication*. Dr. Kellogg was convinced that virtually all illnesses originated in the bowels and that the "putrefaction changes which recur in the undigested residues of flesh foods" were to be blamed for disease. The "cure" for autointoxication was simple: the bowels had to be cleansed! And Dr. Kellogg knew exactly how to do this. First, copious amounts of water were used to flush the bowels through the rear portal. This delightful procedure was then followed by the "yogurt treatment" from both ends. Dr. Kellogg believed that the bacteria used to make yogurt protected against disease and "should be planted where they are most needed and may render the most effective service. . . . Balance your intestinal flora," he maintained, "and you'll live as long as the rugged mountain men of Bulgaria!" And according to Elie Metchnikoff, the Russian bacteriologist whose research triggered Kellogg's yogurt compulsion, that was pretty long.

Metchnikoff had caused quite a sensation with his theory that the longevity of Bulgarians was due to the copious amounts of yogurt they ate. He even had a theory to explain how this happened. The good bugs, *Bacillus bulgaricus*, which Metchnikoff named in honor of the Bulgarians, overwhelmed the bad bugs that caused disease in the gut. Metchnikoff didn't really have any evidence for his theory, or indeed for his notion that Bulgarians experienced remarkable longevity. But when he received the Nobel Prize in physiology and medicine in 1908 (for work unrelated to yogurt), the word about the supposed miraculous properties of yogurt began to spread. And it has been spreading ever since, as the notion of introducing beneficial bacteria into the gut receives more and more scientific support. It is becoming increasingly clear that the staggering number of bacteria that are found in our digestive tract play an important role in health and disease. Their numbers are in the trillions, outnumbering the cells in our body by a factor of ten. Little wonder then that research into *probiotics* is mushrooming.

What is a probiotic? The term refers to any food, beverage, or dietary supplement that contains specific microorganisms in sufficient numbers to alter the microbial flora in a host and exert beneficial health effects. The idea is that these "good" bacteria will multiply and compete for nourishment with the nasty microbes capable of causing disease. Eventually, the theory goes, the nasties will starve and their numbers will wane. It may be surprising to see the adjective "good" associated with bacteria, but it is indeed merited. Some produce enzymes that help digest food, others can synthesize vitamin K in the gut or even help stimulate the immune system. On the other side of the ledger, we have microbes such as *Helicobacter pylori*, the culprit that can cause ulcers. Probiotic therapy can therefore be simplified as "in with the good, out with the bad."

Traditionally, yogurt has been made with *Lactobacillus bulgaricus* and *Streptococcus thermophilus*. These bacteria, though, are acid sensitive and do not make it through the stomach to the colon in sufficient numbers to alter the flora there. On the other hand, *acidophilus* and *bifidobacteria* are more stable to acid and do survive the journey. And when they set up shop in the intestine, they really do squeeze out disease-causing bacteria, such as those responsible for diarrhea. Taking antibiotics for routine infections often causes diarrhea, as some of the good microbes are indiscriminately wiped out along with the disease-causing bugs. Probiotics can replenish the desirable microbes and curb the diarrhea. But that may be just the beginning as far as probiotic benefits are concerned. Tantalizing evidence exists that probiotics may play a role in preventing cancer, enhancing the immune system, and alleviating the symptoms associated with ulcerative colitis and irritable bowel syndrome (IBS). Some probiotics can destroy cancer-causing agents in the gut, and at least one excellent study has shown that the risk of eczema in babies can be reduced if they are given *Lactobacillus* GG. Chances are that this will work for some allergies as well. Particularly noteworthy is the fact that in over 150 studies of probiotics, no adverse effects have been seen.

The vexing question now is to determine which probiotic bacteria are the most beneficial and how sufficient numbers can best be introduced into the appropriate place in the digestive tract. *Lactobacillus* GG (so named by its discoverers, Dr. Sherwood Gorbach and Dr. Barry Goldin) looks very promising. It performs well against diarrhea, shows anticancer effects in animals, and in some cases has even relieved the symptoms of ulcerative colitis. So has VSL#3, a research mix of eight bacterial species. Bio-K+ is a commercially available product that in clinical studies has been shown to deliver the goods, namely viable organisms in sufficient numbers to the colon. On the other hand, there are products out there that claim to contain a host of beneficial bacteria but in fact do not. There are no regulations about ensuring that the number of bacteria claimed on the label are actually in the product. In general, probiotics should contain at least one billion viable organisms per serving to have a chance at being effective, but independent analysis shows that many products fall short.

Manufacturers are leaping onto the probiotic bandwagon, producing yogurts with a variety of desirable microbes. *Lactobacillus casei* is added by some to "improve immunity" and *Bifidobacterium animalis* by others for "digestive health," but the claims are not backed up with solid evidence. Still, yogurt is unquestionably a healthy food and those that contain "live and active cultures" may play a role in improved intestinal health.

And who could have ever imagined that bacteria in the gut could play a role in weight control? But this may indeed be the case. Dr.

Jeffrey Gordon and his team at Washington University School of Medicine in St. Louis may have found an explanation for a mystery that has puzzled many scientists. How is it that two people can have diets essentially equal in calorie content, engage in physical activities to the same extent, and yet have different propensities for weight gain? How can one easily maintain body weight while the other constantly struggles? It seems the answer may lie in the type of bacteria that inhabit their respective guts.

What we are talking about here are some of the "friendly" bacteria that help us digest food by breaking down complex carbohydrates in grains, fruits, and vegetables into simple sugars that can be used for energy. Not all bacteria, however, are equally adept at doing this. The carbohydrate-digesting bacteria fall into two broad classes, the firmicutes and the bacteroidetes, with the latter being less efficient at breaking down carbs. The result is that a high bacteroidetes population leads to more of the complex carbohydrates being excreted, meaning a reduced likelihood of weight gain. By contrast, if firmicutes dominate, the polysaccharides are broken down to simple sugars that are absorbed into the bloodstream. If these are not used for energy, they will be converted to fat and stored in the body.

Supporting evidence for the role of bacteria in weight control comes from studies in mice as well as in humans. Obese mice, for example, have been found to have a lower proportion of bacteroidetes bacteria, but even more intriguing is the finding that when obese people are placed on a low-calorie diet, the levels of bacteroidetes in their gut increase as their weight decreases. Perhaps an imbalance in gut bacteria makes some people prone to obesity and maybe alteration of levels might lead to a treatment. Maybe sometime in the future overweight people will be able to avail themselves of a probiotic to help them lose weight.

Consuming active cultures may not be the only way to increase the numbers of friendly bacteria in the gut. Prebiotic treatment may be an alternative approach. Prebiotics are substances, such as fructo-oligosaccharides (FOS), lactulose, or inulin, that can be included in the diet to stimulate the growth of specific bacteria in the colon. These complex carbohydrates are just "fiber," meaning that they are not digested as food. They pass through the stomach and small intestine unchanged and collect in the colon where they serve as tasty morsels for friendly bacteria. These bacteria then multiply and crowd out worrisome microbes.

In Japan, numerous foods fortified with fructooligosaccharides and inulin are already on the market, and the trend is coming our way. Where do these chemicals come from? They occur naturally in onions, garlic, and bananas, but not to an extent that would have a significant effect on colonic bacterial populations. A daily dose of at least four grams of prebiotics is needed for any hope of benefit, but double that amount is preferable. Just about the only way such an intake can be achieved is by adding FOS or inulin to processed foods. The most common source of the chemicals is chicory root, from which they can be readily extracted.

One plant that does contain a significant amount of these prebiotics is the Jerusalem artichoke. Samuel de Champlain first learned about this tuber from the Native Americans and introduced it to Europe. Actually, it is not an artichoke and has nothing to do with Jerusalem. The plant is a member of the sunflower family and is sometimes called "sunchoke." But it seems that to Champlain it tasted like an artichoke and the term stuck. Why Jerusalem? When the plants were first brought back to Italy from America, they were called *girasole* for "turning to the sun." Somehow this got corrupted to Jerusalem. In Europe and Japan, Jerusalem artichoke flour is already being added to foods to improve their health potential. There may be a downside to this tuber though, as first voiced by John Goodyear, a British farmer in the 1860s. "In my judgement, which ever way they be drest and eaten, they stir up and cause a filthie loathsome stinking winde with the bodie, thereby causing the belly to be much pained and tormented, and are more fit for swine than for men." He may have been right about the wind, but was surely wrong about the Jerusalem artichoke not being fit for humans.

BOOSTING IMMUNITY WITH GLUTATHIONE

The more you know about the workings of the human body, the more amazing it becomes that anyone is ever healthy. Just take a moment to think of all the processes that have to go right to keep us on an even keel. For a start, amino acids have to link together to make proteins, neurotransmitters have to be synthesized, red blood cells have to make hemoglobin, white blood cells have to produce antibodies, and various glands have to secrete hormones. And enough energy has to be generated to ensure that our hearts beat, our lungs breathe, our cells divide, and our brains think. Accomplishing all of these processes requires the concerted action of numerous chemical reactions, and everything has to happen in the face of a constant assault from bacteria, viruses, fungi, and environmental toxins, both natural and synthetic. As if that were not enough, various *reactive oxygen species*, which form as by-products of the reactions needed to sustain life, are constantly poised to shorten it. So, thank goodness for our immune system, that highly specialized collection of body structures and cells whose mission is to recognize and eliminate potential threats to our health.

We know, of course, that the immune system is fallible. After all, people do succumb to the effects of bacteria, viruses, and cancer cells, especially as they get older. Devastating diseases such as AIDS can cause the immune system to virtually self-destruct. Obviously, then, any intervention that boosts the immune activity is most welcome. One such possibility is to raise the levels of glutathione inside of cells.

I suspect that glutathione is not a word that commonly comes up in conversation, unless that conversation is between scientists discussing the immune system. Then you'll hear "glutathione" bandied about with great enthusiasm, because this relatively simple molecule is involved in a number of critical reactions needed to maintain health.

Let's start with glutathione's role in helping white blood cells fight off foreign intruders such as viruses or bacteria by forming protective molecules called antibodies. In order to crank out enough antibodies, white blood cells have to multiply rapidly. This process requires a great deal of energy, which is provided by the reaction of stored nutrients with oxygen. Unfortunately, this reaction also results in the formation of by-products known as reactive oxygen species, a family of free radicals capable of wreaking havoc with the cell's molecular machinery and slowing the immune response. One of the most reactive oxygen species is known as superoxide. Intense investigation of this substance began in 1968 when researchers isolated an enzyme, superoxide dismutase, which had the ability to destroy superoxide by converting it to oxygen and hydrogen peroxide.

But this was short-term protection. It turned out that hydrogen peroxide itself could go on to generate extremely reactive hydroxyl free radicals, capable of causing extensive cellular damage. As we soon learned, though, our fascinating immune system had a way to deal with this problem as well. Two other enzymes, catalase and, more importantly, glutathione peroxidase, effectively eliminated hydrogen peroxide and thus afforded protection against the ravages of what we call *oxidative stress*. As the name implies, glutathione peroxidase uses glutathione to eradicate hydrogen peroxide. Could increased cellular levels of glutathione then help the enzyme do a more effective job and improve immune function?

As scientists were pondering this question, some other interesting properties of glutathione came to light. The molecule itself had antioxidant properties and destroyed free radicals, independent of the glutathione peroxidase connection. It also enhanced the activity of vitamin C. And glutathione had yet another trick up its sleeve. When attached to a toxin (by yet another enzyme, glutathione S-transferase), glutathione rendered the toxin water soluble and excretable. Given all of this, it stands to reason that higher levels of glutathione in the blood should correlate with good health. And that does seem to be the case, at least if we go by a provocative study carried out at the University of Birmingham in England.

Researchers measured glutathione levels in healthy volunteers, both old and young, as well as in elderly patients who either suffered from chronic diseases or had recently been admitted to the hospital with an acute problem. If glutathione were indeed related to health, such an analysis should muster evidence, the scientists suggested. And they were not disappointed. The healthy, young volunteers had the highest plasma levels of glutathione, followed by the healthy elderly, then the elderly outpatients, and finally the elderly inpatients. At the School of Public Health, University of Michigan, Dr. Mara Julius and colleagues also found that higher glutathione levels were associated with fewer illnesses such as arthritis, diabetes, and heart disease. There is also some pretty interesting data from animal experiments about the chemoprotective properties of glutathione. Rats exposed to aflatoxin, a potent carcinogen, fared significantly better when given glutathione. The animals not treated with glutathione all died within two years of exposure to the carcinogen, but 80 percent of the treated animals were still alive at the end of that period.

So the question is, why are we not wolfing down glutathione supplements? Since glutathione can be prepared in large amounts by fermentation methods using specialized yeasts, such supplements are readily available. There certainly is no safety issue. Nobody has ever shown any hazard associated with oral glutathione. The problem is that nobody has shown any major health benefit either. How come? Because unlike in rats, in humans glutathione is not absorbed into the blood from the small intestine to any significant extent. That doesn't mean oral glutathione is completely useless. It can be used for benefit by cells that line the intestine. These cells often become damaged in diseases such as AIDS or cancer, resulting in impaired nutrient absorption, which in turn leads to loss of weight and muscle mass. Oral glutathione has been shown to repair intestinal cells and has been used to advantage in AIDS patients. But for other health benefits, oral glutathione won't do.

We know that our cells are capable of making glutathione, so why not exploit this ability by giving them plenty of raw material to work with? Why not just increase our dietary intake of glutamic acid, glycine, and cysteine? Our food supply actually contains plenty of glutamic acid and glycine, so we don't need to be concerned about these. But cysteine is less abundant, and therefore it is the determining factor in how much glutathione forms. It's sort of like making bicycles. Each one needs two wheels and a frame; it doesn't do any good to have a huge excess of wheels, you cannot make more bicycles than the number of frames available. The frames then are the limiting component, just as cysteine is in the synthesis of glutathione. So why not just start swallowing cysteine capsules? The substance can be readily made by the decomposition of proteins in human hair, a process carried out by several companies in China, mostly aimed at providing cysteine for the food and personal-care industries. Cysteine is used in the production of artificial meat flavors, as a dough-conditioning agent, and in hair-curling products.

Unfortunately, taking cysteine as a dietary supplement does not appear to be a great option. Some animal studies have suggested that it can increase triglycerides and cholesterol levels, and that it can even have neurotoxic effects. People have also reported experiencing nausea. Furthermore, cysteine is not very soluble, and it can also undergo a variety of reactions in the bloodstream, making it unavailable for absorption by cells. There are, however, ways to circumvent this problem. Cysteine can be readily converted in the laboratory into N-acetylcysteine (NAC), which is more soluble and less prone to destruction in the bloodstream. After being absorbed into cells it is reconverted into cysteine, which then is available for glutathione synthesis.

The multitudes of people who have been brought back from the brink of death after overdosing on acetaminophen can attest to the efficacy of NAC. Acetaminophen (Tylenol is a common trade name) is a widely used over-the-counter pain reliever. In the recommended dosage it is very effective, but like any medication it becomes a problem when abused. Overdosing on acetaminophen, especially together with drinking too much alcohol, can lead to severe liver damage and possibly death. Suicide attempts with acetaminophen remain only "attempts" because of quick medical intervention with NAC. Our body recognizes acetaminophen as an intruder and tries to eliminate it by converting it into a more soluble compound. Unfortunately, this compound, N-acetyl-p-benzoquinone imine (NAPQI), is the one that is toxic to the liver; however, a glutathione-induced enzyme stands ready to help the kidneys excrete it. When the acetaminophen dose is too high, glutathione stores are depleted and liver damage ensues. Unless, that is, NAC is quickly administered to replenish the cellular levels of glutathione. Glutathione enhancement with NAC is one of the most effective medical treatments in existence.

If NAC works so well to boost glutathione, why are we not being urged to take it in supplement form to prevent disease? We *are* being urged to do so . . . by manufacturers of NAC. While there is no major toxicity concern here, nausea is a possible side effect. And furthermore, nobody has really examined the consequences of taking NAC over the long term, or its possible interactions with other medications.

It would be great if an innocuous way of increasing our glutathione levels were available—and maybe it is. Cheese-making requires the separation of the curds from the whey, a watery liquid residue that harbors proteins rich in cysteine. When processed in a special way, these proteins can deliver their cysteine content to cells, where it can be released to enhance glutathione formation. Some intriguing studies have already indicated that these special whey preparations have increased endurance in athletes, presumably by reducing free-radical damage to muscles. There's more. Laboratory work has shown increased glutathione levels in human prostate cells exposed to whey protein, suggestive of protection against prostate cancer. Animals fed the protein concentrate are more resistant to carcinogens and, in a somewhat curious finding, whey has been seen to increase glutathione in normal cells, while it depletes it in cancer cells, making the latter more susceptible to eradication by chemotherapy or radiation. Studies are currently under way to see if daily intake of the cysteine-supplying whey protein is helpful in reducing the levels of environmental toxins in the blood. That's the kind of evidence we need before jumping on the bandwagon with both feet, but as far as glutathione supplementation goes, we may have found a "whey" to go.

ADDING FLUORIDE TO WATER

What is the most common disease in the world? Infections probably come to mind. Or heart disease, or cancer, or perhaps even AIDS. Actually, it is the common cold! And what comes in second? Tooth decay! Cavities are unsightly and can cause pain but, more importantly, poor oral health can also allow bacteria to enter the bloodstream and precipitate respiratory or heart problems. Luckily, though, tooth decay is preventable. Proper oral hygiene and reducing sweets in the diet are essential, but it is also possible to make teeth more resistant to decay by chemical intervention—that is, by using fluoride.

Tooth enamel is mostly composed of hydroxyapatite, a substance readily damaged by acids. Naturally occurring bacteria in the mouth, *Streptococcus mutans* in particular, feed on sugars, metabolize these to acids and, presto, cavities appear. However, if fluoride is supplied in the diet or is applied to teeth topically, it gets incorporated into the structure of the tooth, forming a more acid-stable substance called fluoroapatite. While it used to be believed that the best way to incorporate fluoride into teeth was to supply it in the diet as teeth were forming, recent research shows that applying fluoride to the teeth topically is a very effective way of preventing cavities. Fluoride has a secondary effect as well. It can interfere with the activity of enzymes, such as the ones bacteria use to convert sugars into acids. Since tooth decay is a universal problem, it would seem that a simple method of prevention, namely the addition of tiny amounts of fluoride to drinking water, would be a welcome solution.

Indeed, many health authorities claim that water fluoridation has been one of the most effective and safest public health interventions ever introduced. But not everyone shares this view. There are those who claim that fluoridation is misguided, ineffective, and risky. It's only done, they say, because governments, industry, and the American military have conspired with public health authorities to change the image of a toxic by-product of the fertilizer industry into a safe tooth-decay preventor. Why? So that it can be disposed of in our drinking water! These culprits, the antifluoridationists claim, have also colluded to keep data about the risks of fluoride from the public and have attempted to undermine the careers of scientists who present opposing views. Controversy is no stranger to science, but it is rare to see an issue that generates—on both sides—as much verbal venom, and as much misrepresentation of the scientific literature, as the flouridation of public water supplies.

First, a little history. In 1901, Frederick McKay opened a dental practice in Colorado Springs, Colorado, and immediately noted that many of his patients had stained or mottled teeth, a condition now known as fluorosis. But he was surprised to find that people who had these unsightly teeth also had very few cavities. The link turned out to be a very high level of fluoride in the Colorado Springs drinking water. McKay's observation then spurred comparisons of the dental health of communities with different amounts of fluoride in the water. When the natural fluoride concentration was greater than one part per million, the incidence of cavities was reduced by between 50 and 65 percent. At this level, only about 10 percent of children showed the faintest signs of fluorosis, and consequently the World Health Organization began to recommend supplemental fluoride where levels were low.

In 1945, Grand Rapids, Michigan, became the first city in the world to adjust its drinking water to a fluoride concentration of one part per million. Brantford, Ontario, followed the same year. In fact, Brantford took part in the first epidemiological survey of fluoride. In 1948 and 1959, the incidence of cavities was compared with that of Sarnia, which had a very low level of fluoride in the water, and with Stratford, which had a concentration of 1.6 parts per million. Sarnia had a high incidence of cavities both in 1948 and 1959, with 90 percent of children between the ages of nine and eleven having cavities, whereas Stratford had a low incidence with only 50 percent of the children affected. In Brantford from 1948 to 1959, the incidence of cavities dropped from 90 percent to 50 percent. As a result, Sarnia introduced water fluoridation. Today, at least in North America, fluoridation is widespread. The American Dental Association is a strong advocate of adding fluoride to water and estimates that every dollar spent on fluoridation saves about fifty dollars in future dental expenses.

From the beginning, fluoridation pushed some people's buttons. Opponents objected to "rat poison" being added to their water supply and claimed that the government had no right to determine the type of water people should drink. Emotions ran high; mistrust flourished. In March 1944, authorities in the Newburgh area of New York State announced that the town's water would be fluoridated. On the day that fluoridation had been scheduled to begin, local health officials were startled to receive complaints about discolored saucepans, digestive troubles, and cracked dentures. In fact, the fluoridation equipment had not been ready on time and no changes had been made to the water!

Water fluoridation *does* reduce the incidence of cavities. The extent to which this happens, though, is the subject of debate. In recent years the ready availability of fluoride toothpastes, mouthwashes, and dietary supplements has reduced the effects attributable to fluoridated water in more affluent areas. Underprivileged communities are the ones that are the most likely to see the benefits of a fluoridated water supply. But at what risk?

Fluoride can be toxic, there is no doubt about it. As the opponents of water fluoridation constantly remind us, it has indeed been used to poison rats. This fact, however, has no bearing on whether or not we should add fluoride to drinking water to improve dental health. Toxicity is always a question of dose. A mouthful of pure sodium fluoride will kill a rat, but the rodent would have to drink roughly 100 liters of fluoridated tap water before suffering the same fate if the water had the usual fluoride concentration of one part per million. And it would have to do this without urinating! Labeling a substance as a "poison" without putting it into the proper context is meaningless and irresponsible. After all, we use "poisons" all the time. The chlorine we use to purify our water can also be used as a chemical weapon. Morphine is an excellent painkiller, but at doses just slightly higher than that needed to alleviate pain, it can put you to sleep— and it doesn't take much more to put you to sleep permanently. Aspirin at a high dose can kill, as can table salt or iron supplements or fluoridated toothpaste. It might be a challenge to do so without vomiting, but in theory one could down a lethal dose of fluoridated toothpaste. Still, this has nothing to do with adding fluoride to water, or indeed, to toothpaste. Neither do the facts that fluoride is used to enrich uranium for nuclear weapons, to prepare Sarin nerve gas, and to isolate aluminum from its ore.

Antifluoridationists also relish pointing out that hydrofluorosilic acid, the chemical commonly used to fluoridate water supplies, is a waste by-product of the fertilizer industry. This is true, but so what? If anything, converting an industrial waste to a useful substance instead of discarding it is highly desirable. These antifluoride arguments are as inane as Senator Joe McCarthy's charge in the 1950s that fluoridation was a communist plot to poison America or, as others alleged, that it was a masterful stroke by the sugar industry to increase sales of sweets without affecting children's teeth. Antifluoridationists actually harm their cause by using such irrelevant arguments and by their excessive fear-mongering. The truth is that there may be legitimate reasons to take a more careful look at the issue.

The major allegations that have been aimed at water fluoridation are as follows: it increases the risk of bone fracture and bone cancer; it may interfere with thyroid function as well as with other biological systems; it may expose the public to contaminants inherent to hydrofluorosilic acid production, and it may cause fluorosis of the teeth. Only the last of these is a clear concern. Dentists do report seeing more teeth with the hallmark white stains of fluorosis in areas where fluoride is added to water. While this is only a cosmetic problem, it is nevertheless a problem. It occurs because the widespread use of fluoridated toothpaste, fluoride mouth rinses, and processed foods and beverages made with fluoridated water has resulted in some segments of the population being exposed to more than the optimal amount of fluoride. It is also quite clear that due to these sources of fluoride, as well as to earlier and better dental care, the gap in incidence of cavities between fluoridated and nonfluoridated areas has narrowed considerably. Although such statistics are hard to confirm, the current incidence of cavities in Vancouver, which has never fluoridated its water, seems about the same as in Toronto, which has added fluoride for more than thirty years.

The other claims against fluoridation are much more nebulous. While laboratory studies and some animal experiments suggest that fluoride can trigger cancer, extensive epidemiological investigations in fluoridated and nonfluoridated communities have shown no difference in cancer rates except for the possibility of a rare type of bone cancer in boys. Fluoride, as can be expected, incorporates into bones as well as into teeth, but some research indicates, surprisingly, that in this case it may lead to a weakening of bones. Again, epidemiological studies have shown that if there is a risk of increased fractures, it is a very small one. Fluoride does interfere with enzyme systems; that is the way it controls bacteria in the mouth. In theory then, it can have a negative effect on various body functions, possibly including thyroid function. But theory is not the same as evidence. Hydrofluorosilic acid, as fluoridation opponents point out, can indeed be contaminated with trace amounts of lead, arsenic, and radium, all of which are undesirable. But the amounts that end up in drinking water from this source are less than what is naturally present in many water systems. It is also interesting to note that tea is a much higher source of fluoride than fluoridated water, but no adverse effects have been linked to its consumption.

The fluoride issue heated up in March of 2006 when the US National Research Council (NRC) released its report entitled *Fluoride in Drinking Water: A Scientific Review of the Environmental Protection Agency's Standards.* Press coverage was extensive, with most articles accurately reporting the overall recommendation that the maximum allowed level of fluoride in drinking water should be reduced from the current four parts per million. But then reporters went on to interpret this recommendation as a call to action about the safety of water fluoridation. This was quite a leap! Let's take a moment to analyze what this report really stated and what conclusions can legitimately be drawn.

Back in 1986, the Environmental Protection Agency, based on the then available evidence, established four parts per million as the maximum contaminant-level goal for fluoride in water, based on the fact that higher concentrations caused weakening of tooth enamel. There was no suggestion that four parts per million was associated with any other hazard, at least not from the EPA. Antifluoridation groups had other views. They claimed that fluoride in water presented a risk for musculoskeletal, neurobehavioral, and endocrine problems and even suggested that it could cause cancer. Numerous studies on all aspects of fluoride in water have appeared since the four parts per million maximum was set in 1986, and the EPA decided it was time to review the evidence to determine if that maximum was still appropriate.

After examining the most recent toxicological, epidemiological, and clinical studies, the expert panel concluded that severe enamel fluorosis can occur in children even at four parts per million of fluoride in water, and that consuming water at this level continuously can lead to weakened bones and increased risk for fractures. Based on this evidence the panel recommended that the four parts per million maximum be reduced. Now for the important point. When fluoride is added to drinking water to prevent cavities, it is added to bring the final concentration to between 0.7 and 1.2 parts per million. Not anywhere near four parts per million! Who then is at risk from drinking water with a four parts per million fluoride content? About one half of 1 percent of North Americans drink water that has a natural fluoride content of four parts per million or more. So this amount of natural fluoride in water is a potential problem, but the National Research Council report said absolutely nothing about any risk from the one (or so) part per million of fluoride added to community water supplies. And the NRC scientists looked at all of the possible health effects, including hormonal issues and cancer. They found no adverse health effects, other than weakened enamel and slight weakening of bones, even at four parts per million! And there was certainly no recommendation to reduce fluoridation below the usual one part per million. We can't conclude that some future study may not raise some other fluoridation issue, but interpreting this NRC report to mean that adding fluoride to water at a level of one part per million presents a risk is plainly wrong.

The risk of fluorosis, essentially a cosmetic problem characterized by faint white lines or streaks on the enamel, is greatest when teeth are erupting. Accordingly, the American Dental Association recommends that infant formulas not be made up with fluoridated water and that fluoridated toothpaste not be used in children under two years of age. Older children should be told to use no more than a pea-sized dollop of fluoridated toothpaste and should be instructed not to swallow any of it.

Current science tells us that water fluoridation is not a likely cause of significant health problems, but that it may no longer be necessary in all communities. Fluoride toothpastes, fluoride treatments by dentists, and fluoride present in food and drinks may be sufficient to prevent dental disease.

SUPPLEMENTING WITH VITAMINS

Vitamins are certainly essential components of the diet and prevent the classic vitamin-deficiency diseases like rickets and scurvy. But some vitamins also have antioxidant properties, which brings up the question of their possible additional benefits. So should we be taking vitamin supplements? A pretty simple question to answer, one might think. After all, there have been literally thousands of studies on how vitamin and mineral intakes relate to health. More than 100 million people in North America believe the question has been answered and take a variety of daily supplements to protect themselves against disease, spending some $25 billion in the process. But could they be on the wrong track?

There are several ways to investigate the potential role of supplements. Surveys can identify those people who take supplements and make correlations to their health status. Alternatively, researchers can measure blood levels of specific antioxidants and relate the findings to disease patterns. Or they can carry out intervention studies in which results are evaluated after subjects take either the substance being tested or a placebo over an extended period. Finally, a "meta-analysis" can be undertaken in which the results of various high-quality studies are pooled to reveal information that is not apparent from looking at single studies.

A typical survey-type, or *observational*, study involved more than 83,000 healthy American physicians who filled out questionnaires about supplement intake and dietary habits. Roughly 30 percent of the doctors regularly took antioxidant vitamin supplements. After about six years, 1,000 or so of all the doctors studied had died of some form of cardiovascular disease. Were the deceased more or less likely to have been taking antioxidant supplements? As it turned out, there was no relationship between supplement intake and cardio-vascular death. Of course, it is possible that physicians are more health conscious than others and pay more attention to their diet, so that they already had a sufficient intake of antioxidants. Some studies have even shown a negative supplement effect. Analysis of data collected from some 70,000 postmenopausal nurses showed that over an eighteen-year period, those who consumed the most vitamin A from food or from supplements had the greatest risk of bone fractures. On the other hand, low vitamin E intake dur-ing pregnancy has been shown to increase the risk of childhood asthma, and women who take vitamin supplements during preg-nancy appear to have a reduced risk of having infants who develop brain tumors.

How about studies of blood levels of vitamins? English research-ers in one case found that among 20,000 people, those who had the highest levels of vitamin C in the blood lived the longest. But was this because of the vitamin C, or was the vitamin C just acting as a marker for increased fruit and vegetable intake? Low levels of folic acid have been linked with breast cancer, heart disease, and, most significantly, with giving birth to babies with neural tube defects. Still, such studies do not show cause and effect. You can never be certain that the observations are not due to some other dietary fac-tor that happens to parallel folic acid in its presence. That's why intervention studies are the most meaningful. And in the case of folic acid in pregnant women, they certainly back up the observational studies. Supplementing the diet with 400 micrograms of folic acid daily significantly reduces the risk of neural tube defects.

It is reasonable to expect that antioxidants such as vitamins E or C, or the vitamin A precursor, beta-carotene, should play a role in pre-venting heart disease. Why? Because it is well known that cholesterol is most likely to damage coronary arteries when it is oxidized, in other words, when its molecular structure is slightly altered by reaction with oxygen. Antioxidants, in theory, should counter this effect. But in practice the story seems to be different. Researchers in Oxford, England, enrolled over 20,000 adults with heart disease risk factors such as diabetes, high blood pressure, or high blood cholesterol in a major study. Half received a daily supplement of 600 IU vitamin E, 250 milligrams vitamin C, and 20 milligrams beta-carotene, while the others got a placebo. The supplements were certainly effective in increasing blood levels of vitamins, as tests clearly showed. But after five years there was absolutely no difference in any form of disease or in death rates between the groups. Maybe, though, these subjects already had the beginnings of cardiovascular disease that could not be reversed with the supplements and perhaps in a healthy group, supplements can prevent disease. Maybe. . . .

As is evident, it is possible to support either side of the "to supplement or not" debate by looking at the scientific literature selectively. But what happens when scientists put all the data together in a meta-analysis? Sometimes they just add to the confusion! That's what Goran Bjelaković and his colleagues at the University of Niš in Serbia and Montenegro apparently did when they examined the relationship between dietary antioxidants and the risk of gastrointestinal cancers. Free radicals can form in the gut and have been implicated in cancer, and fruits and vegetables have been shown to be protective, presumably because of their antioxidant content. So it certainly seemed reasonable to expect that antioxidant supplements should be beneficial in preventing cancer. Bjelaković scoured the scientific literature and identified fourteen rigorous, placebo-controlled trials involving more than 170,000 subjects. All the trials used oral supplements, although amounts varied, as did combinations. Vitamin C ranged from 120 to 2,000 milligrams a day, vitamin A from 1.5 to 15 milligrams, beta-carotene from 15 to 50 milligrams, selenium from 50 to 228 micrograms, and vitamin E from 30 to 600 IU. The supplements were taken for years, either daily or every other day. Such doses are typical of what average consumers might take.

The results of the meta-analysis were unexpected. No protection against esophageal, gastric, colorectal, pancreatic, or liver cancer was found. Selenium supplementation in a few of the trials did show some optimistic results. Now for the real shocker: in seven trials, all of high quality, involving over 130,000 subjects, the supplement takers had a higher rate of premature death! The researchers actually calculated that one premature death would be expected for every 100 people taking supplements. Little wonder that this work prompted sensational headlines such as "Vitamins Only Take You Closer To Death." How do we interpret this surprising finding? The study was well executed and has statistical weight, but is it not possible that people who are ill are more likely to take supplements and that this explains the increased mortality? Or that supplements are most effective when taken for longer periods? And maybe they don't protect against cancer but do have other benefits.

Dr. Bjelaković decided to look into this last possibility by mounting a second meta-analysis. His team tracked hundreds of published trials on the health effects of beta-carotene, vitamin A, vitamin C, vitamin E, and selenium supplements, and whittled these down to sixty-eight that met the criteria for proper blinded, randomized, placebo-controlled trials. Some studies used low doses of supplements, some high; some lasted months, others many years. Some used single antioxidants, others used various combinations. But the strength of a meta-analysis lies in pooling results from many studies, evening out variables, and allowing an overall conclusion to emerge. As in his previous study, Bjelaković found no benefit from the supplements, and as before, he noted an increase in mortality among supplement takers. The data appear to be robust. More than 230,000 participants were involved in the sixty-eight trials, twenty-one of which focused on healthy subjects who were taking antioxidants to prevent disease.

No surprise, the salvo of criticism has been furious. Many relevant trials were excluded, critics claim. Causes of death were not determined and may have had nothing to do with supplements. Subjects took numerous other supplements and prescription drugs that could have clouded the issue. There is no biological mechanism that can explain potential harm by antioxidants. Well, that one is not exactly true. Certain white blood cells, for example, attack toxins by generating free radicals, and it is possible that antioxidants can interfere with this activity. No doubt, some of the criticism aimed at the Bjelaković analysis is valid, but given that so many subjects and so many studies were involved, any significant benefit from the antioxidants examined would have become apparent. Incidentally, Dr. Bjelaković and his group received no funding from any commercial enterprise and appear to have no reason to either knock or support dietary supplements.

While I do not think that vitamin supplements are killing us, there is mounting evidence that it is better to get our vitamins from food than from pills. It seems that there is an almost magical blend of antioxidants, minerals, and probably other unrecognized ingredients in fruits, vegetables, and whole grains that cannot be replicated in supplements. A thirteen-member expert panel of the National Institutes of Health in the United States concluded that there is insufficient evidence for *or* against recommending vitamin supplements except in three cases. Supplementation with B vitamins in women of child-bearing age is beneficial, as is supplementation with calcium and vitamin D in postmenopausal women to prevent bone fractures. And the progression of macular degeneration can be reduced with a mix of beta-carotene, zinc, vitamin C, and vitamin E. Note that taking supplements to prevent heart disease is not one of the recommendations. That may surprise many people because it has become almost dogma that certain vitamin and mineral supplements can prevent hardening of the arteries (atherosclerosis), which can lead to heart disease.

There is no doubt that, in the test tube, antioxidants such as vitamins E and C, the vitamin A precursor beta-carotene, and the mineral selenium can reduce the free-radical damage that plays a role in the development of atherosclerosis. The B vitamins also have been presumed to have a protective effect because they are an important factor in lowering the levels of homocysteine in the blood, an independent cardiovascular disease risk factor. As we have already seen, studies of human populations have shown that higher concentrations of homocysteine are associated with an increased risk of heart disease. And many researchers have noted that populations with a low dietary intake of antioxidant vitamins show greater progression of atherosclerosis. Such observations, however, cannot prove cause and effect. People who have low antioxidant intake probably have numerous other lifestyle differences as well. Proving cause and effect requires randomized, controlled trials in which supplements are given to one group, and a placebo to another. So far, as we have seen, such clinical trials have not shown a major protective effect in terms of preventing the symptoms of heart disease. But there is always the lingering possibility that the trials have not been long enough.

That is exactly why researchers at Johns Hopkins Hospital in Baltimore decided to investigate whether supplements can slow the process leading to atherosclerosis, by peeking directly into the human body. These days there are a number of imaging techinques, including angiograms, ultrasound, MRI, and CAT scans that can actually document the extent of hardening of the arteries. Dr. Eliseo Guallar and colleagues identified eleven randomized, controlled trials that involved giving patients antioxidant supplements or B vitamins and that monitored the status of their coronary arteries. Two of the antioxidant trials used only vitamin E; three used a combination of vitamins E and C; and the others used various combinations of vitamins E and C, beta-carotene, and selenium. Several trials used only the B vitamins. In other words, all of the supplement methodologies that had been promoted to reduce cardiac risk were explored. The results were very disappointing. None of the vitamin combinations reduced the progression of atherosclerosis. Furthermore, the vitamins had no effect on preventing the closure of coronary arteries that had been opened up by balloon angioplasty. Based on the thorough evaluation of these well-controlled studies, the Johns Hopkins researchers concluded that the widespread use of vitamin-mineral supplements to prevent atherosclerosis is not supported by the scientific evidence.

In spite of a lack of scientific support, many people take supplements as "nutritional insurance," just in case their diet is inadequate. There is no great risk here and possibly even some benefit as long as megadoses are avoided. This caveat is underlined by a National Cancer Institute study published in 2007 that showed a link between excessive use of multivitamin supplements (more than one multivitamin a day) and an increased risk of advanced prostate cancer. There was no problem with men who took just one daily multivitamin; in fact, there seemed to be a slightly protective effect against the disease. So where does this leave us? The scientific consensus is that vitamin C in the 250 to 500 milligram range is safe enough, as is vitamin E in doses up to 400 IU. Vitamin A should not exceed 4,000 IU, and it is preferable if some of it comes from beta-carotene, its precursor. The best case for supplementation can be made for vitamin D and the B vitamins, particularly folic acid. We have seen the tantalizing data linking vitamin D with protection against various cancers and the studies that suggest the risk of dementia is reduced with adequate B vitamin intake. Supplements containing about 2 milligrams of vitamin B_6, 6 micrograms of B_{12}, and 400 micrograms of folic acid can compensate for a lack of these in the diet. As far as vitamin D goes, many researchers now believe that we should be getting about 1,000 IU a day, an amount that is hard to achieve without supplements.

Although vitamin purveyors often clamor about their products being of higher quality than others, differences between the major brands have no practical significance. Often, the same manufacturer produces vitamins for various distributors, which are then sold at varying prices. While the actual value of taking vitamin supplements is questionable, there is no doubt that for many people they provide comfort and hope, which may be valuable at any price.

MANIPULATING GENES IN OUR FOOD

The ancient Greeks did not have a good grasp of genetics. A giraffe, they thought, was a cross between a camel and a leopard, and an ostrich was the result when a camel mated with a sparrow. A tough task for the bird, one would think. Why did they hold such beliefs? Because in the absence of facts, imagination stepped in. And it still does today. A recent survey showed that one-third of all Europeans believe that tomatoes contain genes only if they have been genetically engineered.

Such surveys are undertaken to gauge public reactions to genetically modified foods, the hottest potato in the area of food safety to come our way probably since the introduction of pasteurization in the early 1900s. Activists back then advised people to spurn the new process because it destroyed the nutritional qualities of the milk and even described the horrors that could arise from consuming "dead bacteria." The truth, of course, was that live bacteria such as *E. coli* and *Salmonella* were the ones worth worrying about. Even today, in the face of all common sense, there are those who promote raw-milk products, framing their resistance to pasteurization as a human rights (freedom to choose) issue.

Today's bogeyman is not pasteurization but genetic modification. Just about everyone has an opinion on the subject, but much too often this opinion is based on hearsay and emotion rather than on scientific data. Consumers speak of "Frankenfoods," and activists attack and destroy experimental fields planted with modified crops while at the same time they clamor for more research into the effects of such crops.

I am not going to suggest that there aren't some contentious issues about genetic modification, just as with any new technology. And I'm certainly not going to say that scientists can absolutely guarantee that genetic modification of foods will have no pitfalls. Nobody can make such a guarantee. Indeed, demanding unqualified assurance about the safety of genetically modified foods is just plain naive. We don't make such demands in other aspects of life. We don't say that we will not fly in an airplane unless we are guaranteed that it will not crash, because we realize that this would be an absurd request. We fly because we know that the benefits outweigh the risks. This is also how we have to look at genetically modified foods.

First of all, let's understand that just because something may be good for Monsanto, Novartis, AstraZeneca or any other company involved in biotechnology, it isn't necessarily bad for the public. If you listen to some alarmists, you can get the impression that these companies are trying to foist poisons on us purely for the sake of profits. But no company wants to undermine its existence or its profits by marketing dangerous substances. A great deal of research has gone into genetic modification and its safety aspects. Many of the potential problems that are now being touted by opponents were in fact addressed long ago by the industry. The testing for allergens in modified foods has been going on since the inception of the technology. In one case, adding a Brazil nut gene to soybeans in order to increase the quality of the protein for improved animal feed resulted in the transfer of an allergen. In other words, someone with a Brazil nut allergy could have reacted to eating the genetically modified soybeans. But the problem was picked up in routine testing, and the soybeans, which had been intended to be used for animal feed only, were never marketed.

We should note that we don't ban peanuts or strawberries or fish because some people have allergies to these foods. And these allergies are far more prevalent than the theoretical allergies to modified foods. Indeed, it may be possible to genetically modify peanuts to eliminate the protein that is responsible for allergies.

Opponents of genetic modification suggest that we should be satisfied with the normal process of cross-breeding plants to produce improved varieties. But where is the guarantee that this procedure doesn't introduce undesired chemicals? Cross-breeding can, for example, yield plants that are more resistant to insects. And why don't insects attack them? Because these plants contain more natural toxins than other plants. Nobody knows the human consequences of eating these natural pesticides. Why are the activists not demanding that all hybrid plants or, indeed, all plant foods be tested for natural toxins?

Genetic modification offers tangible benefits. Combatting malnutrition, for one. When people think of malnutrition, they usually think of starving children. But that is not the only kind of malnutrition in the world today. In fact, the most common kind of malnutrition is iron deficiency, which can cause intellectual impairment, suppressed immunity, and complications in pregnancy. Millions of people in the world suffer from iron-deficiency anemia. Most subsist on rice as their dietary staple, a grain that contains very little iron, and the iron it does contain is unabsorbable because of the presence of substances called phytates. These compounds bind iron in the digestive tract and can prevent it from being transported across the intestinal wall into the bloodstream.

Genetic modification has resulted in a variety of rice that has more iron. This was accomplished by inserting a gene isolated from French beans (also called kidney beans, flageolets, and haricot beans) into the DNA of the rice. This particular gene codes for the synthesis of a protein called ferritin, which is an iron-storage protein. In other words, the rice now can incorporate more iron from the soil. Furthermore, another gene, this time from a fungus, that codes for an enzyme that breaks down phytates was also incorporated, thereby making iron more available.

Populations that subsist on rice also suffer from vitamin A deficiency. That's because rice is very low in beta-carotene, the body's precursor for vitamin A. Deficiency of this vitamin is a major cause of blindness in the developing world; it is estimated that some 250 million children have vitamin A levels low enough to cause impaired vision. Lack of vitamin A also predisposes to various cancers and skin problems.

Vitamin A deficiency was addressed by introducing into rice flour genes that code for proteins that enhance beta-carotene synthesis: two from daffodils and two from a bacterium. The rice is yellow, clearly demonstrating that it is now fortified with beta-carotene. Experiments are under way to cross the iron-rich rice with the beta-carotene-rich rice to produce a variety of "super rice" that can alleviate nutritional problems that affect billions of people.

There are many other fascinating possibilities. How about genetically modifying foods to contain higher levels of cancer-fighting compounds such as sulforaphane found in broccoli? Or fresh fruits and vegetables with improved shelf lives? Edible vaccines? Crops that will flourish in salty soil? All realistic possibilities.

But I can hear the critics' minds churning away. Why am I not talking about monarch butterflies being killed by corn that has been engineered to contain a gene from the *Bacillus thuringiensis* (Bt) bacterium to protect it against the European corn borer? Or the possibility of weeds developing resistance through cross-pollination from crops that have been genetically engineered to be herbicide resistant? Or a study that claimed rats fed genetically modified potatoes developed gastric problems? Simply because, in my judgment, based on the available scientific literature, these concerns have been addressed and found to be either nonexistent or solvable. Surrounding a cornfield with a few rows of non-Bt corn, for example, minimizes the monarch butterfly problem.

Genetic modification is a hugely complex scientific, economic, political, and emotional issue. It is possible that those who support it may have to eat crow at some point in the future, if it is proven to be harmful. But by then we'll probably have a genetically modified version that is nutrient filled and highly palatable.

FARMING ORGANICALLY

s it a fruit or a vegetable? That used to be the major tomato dilemma. Not any more. Now people want to know if a tomato was grown organically or conventionally. They speculate about a tomato's lycopene content. They wonder about the relative nutritional merits of cooked versus raw tomatoes. How did eating one of nature's most delicious foods get so complicated?

One of the tomato's key nutrients is lycopene. Lycopene not only gives color to tomatoes, as well as pink grapefruit and watermelon, but it has another property as well. Lycopene is an antioxidant, meaning that it can neutralize free radicals. A number of studies have suggested that a diet containing lycopene may offer protection against cardiovascular disease and macular degeneration, as well as against cancer of the prostate, the cervix, and the gastrointestinal tract. Although the evidence is not conclusive, there is certainly no harm in increasing our lycopene intake. Wouldn't it then be fruitful to know which tomatoes have the highest levels of lycopene, and while we're at it, the highest levels of other antioxidants such as beta-carotene, vitamin C, and the polyphenols?

This is not a simple question to answer. The nutritional composition of produce is affected by many factors, including sunlight exposure, moisture, type and amount of fertilizer used, extent of attack by pests, and plant genetics. Red tomatoes, for example, can have three times as much lycopene as pink tomatoes (you can forget about lycopene in fried green tomatoes). Red cherry tomatoes have more lycopene per gram than large red tomatoes and also have more polyphenols. Then there are variations depending on the type of tomato, whether it is field grown or greenhouse grown, and its degree of ripeness when picked. And what about organic tomatoes, grown without the use of synthetic pesticides or fertilizers? Are they more nutritious?

When French researchers compared the differences in lycopene, vitamin C, and polyphenol content of organic versus conventional tomatoes, they found that the organic tomatoes had somewhat higher levels of vitamin C and polyphenols, which was not surprising given that the tomatoes probably produce these to fend off pests. If plants get no help from commercial pesticides, they will produce more of the natural variety. Lycopene levels did not differ between organic and conventional tomatoes. Furthermore, the researchers investigated blood levels of these substances in people fed ninety-six grams daily of either organic or conventional tomato purée for three weeks and found no difference in lycopene, vitamin C, or polyphenol levels.

A fascinating study carried out in Taiwan matched ten conventional and ten organic tomato farms and found that there was no difference in the lycopene, beta-carotene, vitamin C, or phenolics content of the produce. Certain farming practices, both in conventional and organic systems, did affect the quality of the tomatoes. Overwatering, for example, reduced lycopene content; weeds reduced carotenoid concentrations; and the phosphorus and iron content of the soil was found to influence vitamin C and phenolic concentrations. On nutritional grounds, whether you eat conventional or organic tomatoes doesn't matter. Taste, however, is another story.

The difference in flavor between biting into one of those giant supermarket tomatoes or into the cardboard box it came in is minimal. That's because over the years we've used various techniques to grow produce faster and to make it bigger. Synthetic fertilizers, with their high levels of nitrogen, potassium, and phosphorus, encourage rapid growth, but their use results in more water being

taken up from the soil. The produce is bigger, but it is bigger because it has a higher water content. Organic crops fertilized with
manure take up nitrogen more slowly and contain less water. In a
sense they are more concentrated in flavorful compounds. And they
contain less pesticide residue, which is another reason that people
gravitate toward organic produce. But is the difference in the
amounts of residue between conventional and organic produce of
practical significance?

One way of coming to some sort of conclusion on this issue is to
compare the Acceptable Daily Intake (ADI) of pesticides as determined by the World Health Organization with the average intake of
these substances in the daily diet. The ADI is determined by first
feeding pesticides to animals to identify the most sensitive species.
Then the highest level of pesticide given on a daily basis throughout this animal's life that does not cause any noticeable toxicological effect is determined. This amount is then divided by a safety factor
of 100 to arrive at the ADI for humans. In other words, a typical
human exposure at 1 percent of the ADI represents an exposure that
is 0.0001 of a dose that causes no toxicity in animals.

In order to determine what the actual human exposure is, the US
Food and Drug Administration used to carry out a Total Diet Study
that involved purchasing 285 different foods typically found in the
diet and analyzing these for pesticide residues. When thirty-eight
of the most commonly used pesticides were examined, thirty-four
were found to be present at less than 1 percent of the ADI, while the
other four were present at less than 5 percent of the ADI. Because
the levels were so low, the FDA has stopped carrying out such a
survey on an annual basis. While the residue from pesticides would
seem to pose very little risk, eating organic foods does eliminate
exposure. When children eating conventional foods are switched to
organic foods, pesticides disappear from the urine after five days.
Of course, the only reason they were detected in the first place is
because our analytical detection capabilities have become so phenomenal that they can find the proverbial needle in the haystack.

Cooked versus raw tomatoes? Lycopene is more readily absorbed from the cooked variety, making tomato sauce and, believe it or not, ketchup, good sources. Interestingly, here *organic* makes a difference, with one study showing organic ketchups having twice as much lycopene as conventional varieties. But remember that you can always double your lycopene intake by eating two tomatoes instead of one. Finally, if you are still wondering, the tomato is indeed a fruit, not a vegetable.

PART THREE

CONTAMINANTS IN OUR FOOD SUPPLY

PESTICIDE CONCERNS

Pesticides are nasty chemicals. They have to be. Sweet smells and pleasant tastes are not going to beat off the myriad insects, weeds, and fungi that look upon our food supply as their food supply. This is a job for poisons. Our challenge is to find ways of using dangerous chemicals safely. It can be done. Today's pesticides are safer and more effective than earlier versions. Whereas a couple of decades ago pesticide application rates were measured in kilograms per hectare, today it is in grams per hectare. The inherent risk of modern pesticides is also less than that of those approved when knowledge of toxicity was far less extensive than it is today.

Let's remember that pesticides were born out of necessity. The cultivation of crops has always featured a relentless battle against pests, a battle that required farmers to take up chemical arms. Thousands of years ago, the Sumerians learned to dust crops with elemental sulphur and the ancient Romans discovered that burning coal tar drove insects from orchards. Later, when the toxicity of lead and arsenic compounds became apparent, farmers began to apply the likes of lead arsenate to their crops, without much concern for effects on human health. Producing enough food to feed the growing population was the prime goal.

Nicotine, pyrethrum, and rotenone extracted from tobacco, chrysanthemum, and derris plants, respectively, had joined the chemical stockpile by the nineteenth century. Malathion and chlorpyrifos, typical organophosphates, were born out of research into poison gases during the Second World War, and the rapid advances in chemistry in the postwar era introduced synthetic pesticides such as DDT, benzene hexachloride, and dieldrin. Insects shuddered, fungi floundered, weeds wilted, and agricultural yields boomed. And, at least in the developed world, worries about lack of food began to be replaced by concerns about pesticides. In the 1960s, Rachel Carson's book *Silent Spring* alerted us to the possible effects of pesticides on biodiversity, and we began to hear faint rumblings of epidemiological studies that linked occupational pesticide exposure to health problems.

Analytical chemists, armed with their gas chromatographs and mass spectrometers, heightened our fears when they found that farmers and agrochemical workers were not the only ones exposed to pesticides. We all were! Residues of these chemicals were found on virtually everything we ate. Apples, for one, were tainted with Alar, a plant growth regulator sprayed on trees to prevent the fruit from falling prematurely. This chemical had cruised under the public radar until 1989, when the popular TV program *60 Minutes* lowered the boom by introducing a segment on Alar with a picture of an apple bedecked with thé classic skull and crossbones, as a reporter enlightened us about the "fact" that "the most potent cancer-causing agent in our food supply is a substance sprayed on apples." People responded by flushing apple juice down the drain and removing apples from children's lunch boxes. The "fact" that Alar was the most potent carcinogen in our food supply was not a fact. True, one of the breakdown products of Alar, 1,1-dimethylhydrazine, did induce tumors when fed to mice in huge doses, an effect that regulators were well aware of when approving Alar for commercial use. The carcinogenicity study highlighted by *60 Minutes* was questionable, they maintained, and irrelevant as a model for human exposure.

Whether or not Alar ever posed a risk is still debated, but there is no doubt that it placed the issue of pesticide residues in food on the front burner. Toxicologists, agronomists, physicians, and environmentalists all waded in with their opinions, along with hordes of emotionally charged consumers who were clearly out of their depth in a such a complex discussion. Bruce Ames of the University of California, one of the most respected biochemists in the world, was quick to point out that we are exposed to all sorts of toxins, both synthetic and natural, on a continual basis, and that more than 99.9 percent by weight of pesticides in the average diet are naturally occurring compounds that plants produce to defend themselves against insects and fungi. Potatoes, for example, synthesize solanine and chaconine, compounds that, like some synthetic pesticides, inhibit the activity of cholinesterase, a crucial enzyme. But we don't shun potatoes because they harbor these natural pesticides. According to Ames and other experts, the body doesn't handle natural pesticides differently from synthetic ones, so there seems to be little justification for all the hand-wringing over remnants of synthetic pesticides in our food supply, usually measured in parts per trillion. Take a football field, pile it with sand to a height of some six meters, mix in one single grain of red sand, and search for it. You'll be searching for one part per trillion.

Some people will argue that there is nothing we can do about the natural toxins, and their presence does not justify a cavalier use of synthetic pesticides. True, but our use of pesticides is anything but cavalier. Regulatory agencies demand rigorous studies before a pesticide is approved. This long and involved process requires acute, short-term, and lifelong toxicology studies in animals, as well as studies of carcinogenicity and possible damage to the nervous system. Proof of absence of birth defects is required. Effects on hormonal changes have to be studied in at least two species, along with the effects of the pesticide on nontarget species. All routes of exposure are assessed, whether via ingestion, inhalation, or skin contact. Cumulative effects are studied. Field testing for environmental effects is also required.

Perhaps the most important facet of approval for consumers is the determination of the maximum dose that causes no effect in a test animal. This dose is then divided by a safety factor of at least

100 to formulate allowable levels for human exposure. And to assess overall risk, the supposition is that the food contains 100 percent of all legal residues and that people eat these foods for seventy years. That such care is taken should sound comforting, especially when we learn that more than 70 percent of fruits and vegetables have no detectable pesticide residues and that only about 1 percent of the time is the legal limit exceeded, a limit that already has a built-in hundredfold safety factor. Nevertheless, we should still wash our produce, although more to remove bacteria than pesticides. A thirty-second rinse significantly reduces both soluble and insoluble pesticide residues. It doesn't eliminate them, not when we can measure residues in billionths of grams.

Just because a substance is found to be present doesn't mean that it presents a risk. Some organizations, such as the Environmental Working Group (EWG) in the United States, are fond of coming up with lists of pesticides found on fruits and vegetables and using these to make recommendations about adjusting eating habits to lower pesticide intake. EWG has pointed an accusing finger at "the dirty dozen" fruits and vegetables most consistently contaminated and claims that people can lower their pesticide exposure by 90 percent by avoiding these and choosing from the "least contaminated" list, which includes corn, avocado, cauliflower, asparagus, onions, peas, and broccoli. Apples, strawberries, raspberries, and spinach are on the "avoid" list despite the fact that they contain a variety of phyto-chemicals known to be beneficial. In any case, saying that one fruit or vegetable is more contaminated than another is meaningless without bringing reference values into the picture. The critical question is whether residues exceed the rigorously determined allowable limits. If not, why the panic? Do we really want to trade in apples for asparagus based on infinitesimal amounts of pesticide residues?

Undoubtedly these debates and those about the validity of using animal models to determine human carcinogenicity, about the existence of a threshold effect for carcinogens, and about trace residues of pesticides that may be harmless individually but not when they are combined, will continue. So will the use of pesticides. By the year 2030, ten billion people will be coming to dinner. But without the sensible use of pesticides they will be going home hungry.

Would a pesticide-free world be better? For people who have to handle pesticides occupationally and for the environment, yes. For the consumer, no. Yields would be significantly reduced, the year-round availability of fresh produce would be limited, and in light of the overwhelming evidence of the ability of fruits and vegetables to protect us against cancer, public health would be compromised.

ACRYLAMIDE IN FRIED AND BAKED FOODS

It all started in 1997 with the unusual behavior of some Swedish cows. Farmers in the Bjäre Peninsula noticed that some of their animals just staggered about, unable to stand up properly. Fish breeders in the area also complained of unusually large numbers of dead fish. It didn't take long for an accusing finger to be pointed at the sealant used to waterproof a tunnel that was being built nearby. Some 1,400 tons of the material had been used, and suspicion immediately fell on the active ingredient, a synthetic polymer called polyacrylamide. The polymer itself is innocuous, but the compound from which it is made, acrylamide, is not. The synthesis of polyacrylamide is based on joining together molecules of acrylamide, much as a chain is made by joining together its component links. But polymerization is never complete, and some residues of the monomer, in this case acrylamide, are always present.

Since acrylamide can show up in drinking water, its toxicity has been extensively studied. The source is polyacrylamide, used in water treatment to coagulate and trap suspended impurities. There is no doubt that when fed in huge doses to test animals, acrylamide can cause a variety of tumors, as well as neurological problems. Accordingly, the World Health Organization has established a maximum dose of 0.5 parts per billion in drinking water, an amount way below the dose seen to cause any effect in test animals. But the concentration of acrylamide in the groundwater around the Swedish tunnel was far greater than this, enough to cause the problems in the fish and cows. This was certainly worrying, but authorities really became concerned when they learned that tunnel workers had also been complaining of numbness in their extremities, a likely sign of acrylamide toxicity.

Margareta Tornquist of Stockholm University was asked to investigate the problem and began by looking at acrylamide exposure among the workers. Blood samples were taken and analyzed for acrylamide content. For comparison purposes, Dr. Tornquist also took some random blood samples from the Swedish population and she got a shock. Sure enough, the tunnel workers had high levels, but so did people who had never been near the area in question. How had they been exposed? The water supply was checked, but no significant amounts of acrylamide were detected. Suspicion then turned to the diet. Analysis of a variety of common foods showed acrylamide to be present in potato chips and french fries, breads, cookies, and crackers. And most terrifyingly, in Swedish crispbread! As was later determined, the source of the acrylamide was a commonly occurring amino acid called asparagine that, in the presence of glucose and high temperatures, undergoes a series of reactions that eventually form acrylamide. Clearly, a carcinogen was being formed from natural components in food—and not in trivial amounts.

The Swedish scientists weren't talking about 0.5 parts per billion, they were finding about 400 parts per billion in french fries, and 1,200 parts per billion in some chips. Far in excess of drinking-water limits! Based on animal data, such levels of acrylamide could in theory result in human cancers. However, we have no evidence that acrylamide is a human carcinogen. A long-term study of more than 8,000 workers who manufacture the substance has not turned up any excess cancers. Let's also be clear that our diet is full of natural carcinogens. Aflatoxins in peanuts, ethanol in wine, urethane in sherry, styrene in cinnamon, and heterocyclic aromatic amines in beef bouillon are as carcinogenic to rodents as is acrylamide. But we don't eat isolated chemicals, we eat food. And food also contains a variety of anticarcinogens. Just think of glucosinolates in broccoli, polyphenols in apples, or lycopene in tomatoes. So while acrylamide may be a carcinogen when fed in large doses to rats, we have no evidence that it causes problems when it is a component of foods. We even have some evidence that it doesn't.

A large case-control study conducted by the Harvard School of Public Health and the Karolinska Institute in Sweden examined the dietary intake of acrylamide among 987 cancer patients and compared this with the diet of 538 healthy people. There was no link between consumption of acrylamide-rich foods and the occurrence of colon, kidney, or bladder cancers. Surprisingly, the study associated higher levels of acrylamide not with a higher, but with a lower, incidence of colon cancer! Perhaps acrylamide-rich foods also contain fiber, which offers protection. An Italian study of more than 7,000 cancer patients came to a similar conclusion, finding no link to acrylamide. Similar results have been found for breast cancer. In rats, high doses of acrylamide do increase the risk of mammary cancer, but a Swedish study of more than 43,000 women found no such link. The women filled out detailed food-frequency questionnaires that allowed researchers to calculate their acrylamide intake. Over eleven years, about 700 of the women were diagnosed with breast cancer but no association with acrylamide was seen.

Food chemists have, however, taken the acrylamide issue to heart and have devised ways to reduce levels in processed foods. Frying at temperatures below 347°F (175°C) significantly reduces acrylamide levels, as does blanching potato chips in dilute acetic acid before frying. When sodium hydrogen carbonate (baking soda) is used to replace ammonium hydrogen carbonate as a leavening agent in baked goods, acrylamide levels are reduced by 60 percent. These measures have proven to be effective, and the estimate now is that we ingest roughly 0.43 micrograms of acrylamide per kilogram of body weight in our diet, an amount well below the dose that can cause cancer in laboratory animals.

Worries about acrylamide cannot and should not be completely dismissed, but producers have done a good job in reducing the amounts in commercial foods and we can do our part to reduce exposure at home by following the "golden rule." When cooking or baking, allow foods to turn a golden color but don't let them go brown or black. And if you are really concerned about acrylamide, you've got to watch your coffee intake. About 30 to 40 percent of our acrylamide exposure comes from drinking that dark brew. But no study has linked coffee to cancer!

ANTIBIOTIC RESIDUES

By and large, drugs don't cure disease. They may lower blood pressure, reduce cholesterol, alleviate pain, restore hormone levels, help control diabetes, or treat erectile dysfunction, but they don't solve the underlying problem. Except for antibiotics. If the diagnosis is bacterial infection, the right antibiotic will be curative. At least for now. But the future for these wonder drugs is more murky. Antibiotic resistance is becoming a huge concern.

Bacteria, like humans, are biochemically unique. Expose a group of people to the cold virus and they will not all come down with a cold. Obviously, the capacity of the immune system to deal with foreign intruders varies from person to person. Similarly, some bacteria can survive the onslaught of antibiotics and then pass their protective genes on to their progeny. The result is a bacterial population that is resistant to the original antibiotic. Such resistance is an inevitable consequence of the use of antibiotics, and the only protection we have against it is the wise use of these powerful drugs. Unfortunately, we have not always been wise.

As pharmaceutical companies successfully developed a wide array of antibiotics, our attitude was that if resistance to one crops up, another will be available to take its place. Until now, this has mostly proven to be so, but the antibiotic cupboard is becoming bare. And there have even been a few chilling reports of resistance to vancomycin, the antibiotic of last resort. Simply stated, the more an antibiotic is used, the less likely that it will maintain its effectiveness. Given that the US Centers for Disease Control estimates that one-third of all antibiotic prescriptions are inappropriate, it is evident that we face a huge problem. Physicians are recognizing this and are becoming less cavalier about prescribing antibiotics. But there is another issue. Although the numbers are somewhat debatable, roughly eleven million of the thirteen million kilograms of antibiotics produced annually in North America are not destined for human use. Instead they are given to hogs, poultry, and cattle, in most cases, not to cure them of disease, but to promote their growth.

Since the late 1940s, so-called subtherapeutic doses of antibiotics have been routinely added to animal feed to prevent disease and to increase feed efficiency. Exactly why animals put on weight more readily when exposed to small doses of antibiotics isn't clear, but it may have to do with reducing the competition for nutrients by cutting down on the natural bacterial population in the animals' guts. Some studies also suggest that antibiotic use thins the intestinal wall and increases nutrient absorption. What has become clear, however, is that such subtherapeutic use of antibiotics leads to the flourishing of antibiotic-resistant bacteria in animals and that such bacteria can infect humans. Chickens, for example, will begin to excrete antibiotic-resistant E. coli in their feces just thirty-six hours after being given tetracycline-laced feed. Within a short time, these bacteria also show up in the feces of farmers. And a truly frightening prospect is that bacteria can pass genes between each other, including the ones that make them resistant to antibiotics. This means that bacteria that have never been exposed to an antibiotic can acquire resistance just by encountering resistant ones. Then consider that animals shed bacteria in their feces and that manure is used as fertilizer, and that fertilizer gets into groundwater, and it quickly becomes evident how the bacterial resistance problem can mushroom.

Thorough cooking kills bacteria, but the widespread incidence of food poisoning demonstrates that poor food handling and undercooking are common. True, most people who come down with bacterial food poisoning just experience some unpleasant cramps and diarrhea and recover without the need for antibiotic treatment. In this situation, resistance is not an issue. But there are numerous cases of children, the elderly, or people whose immune system is compromised, who need antibiotic treatment for food poisoning. If the bacteria are resistant to antibiotics, these patients can face a dire situation. Take, for example, the case of an unfortunate Danish woman who died in 1998 after eating *Salmonella*-infected pork. She failed to respond to ciprofloxacin (Cipro), the usual antibiotic of choice, because of bacterial resistance. In a piece of elegant research, Danish scientists succeeded in genetically matching the *Salmonella*-resistant strain to a specific pig farm. Surprisingly, these pigs had not been treated with ciprofloxacin, but the pigs on neighboring farms had been, and the resistant bacteria had moved between farms!

In North America antibiotics known as quinolones have been used since 1995 to treat infections in poultry. While this was great for the chickens' health, it turned out not to be so good for humans. The most common cause of bacterial gastroenteritis in people is *Campylobacter jejuni*, and poultry is often responsible. If an antibiotic is needed, ciprofloxacin is the usual choice. But since the introduction of quinolones to farm animals, *Campylobacter* strains resistant to the drug have emerged. The US Food and Drug Administration has recognized this as such a serious problem that it has made Baytril, a quinolone, the first veterinary drug to be banned because of the emergence of resistant bacteria. While this is the first action of its kind in North America, Europeans have been phasing out antibiotics in animal feed since the 1980s. Sweden banned the use of antibiotics as growth promoters in 1986, and Swedish farmers responded by improving hygiene on farms and by altering feed composition. They showed that meat can be produced for the consumer at virtually the same cost as with antibiotics. The European Union followed suit and on January 1, 2006, banned the use of antibiotics as growth promoters in animal feed.

Antibiotics are wonderful drugs and we must do all we can to protect their efficacy. While certain uses of antibiotics to treat sick animals are justified, as one scientist who studies antibiotic resistance opined, "Cipro is an essential antibiotic, and we cannot allow its effectiveness to be compromised by squandering it on poultry."

HORMONES IN MEAT

Do the Europeans know something we don't? They banned the use of hormones as growth promoters in cattle in the late 1980s, but the practice is still widely followed in North America. What's going on? How can two continents with some of the best scientists in the world come to different conclusions based on the same scientific evidence? Maybe because the evidence is not conclusive, or maybe because more than science is involved.

There is no argument over the issue that growth promoters work, at least as far as cattle producers are concerned. Steroidal hormones added to feed, or implanted into the ears of animals, increase growth by about 20 percent and allow farmers to use 15 percent less feed than in untreated cows. This practice translates to lower consumer prices, as is evident to anyone who has purchased meat on both sides of the Atlantic—but at what cost to human health?

Our hormone story takes us back to 1938, when Charles Dodd in Britain first synthesized a compound that mimicked natural estrogen. Diethylstilbesterol (DES), which was easy and cheap to produce, garnered immediate attention. It could be taken orally and offered hope to women in preventing miscarriage, as well as in treating menstrual problems, menopausal symptoms, and morning sickness. What excited farmers, though, was the effect that DES had on animals. Poultry and livestock gained weight more quickly when the compound was added to feed. Given that DES had already been approved for medical use in humans, approval as a feed additive in

1954 generated little concern. But it didn't take long for some anxieties to arise. There was talk that male agricultural workers exposed to DES experienced breast growth, and there were murmurings that DES in poultry was triggering precocious puberty in girls. Although this was never confirmed, DES was banned in poultry and lamb production in 1959. Use in cattle continued, even after the hormone was linked to a rare form of vaginal cancer in the daughters of women who had taken it during pregnancy. Because of this connection to cancer, DES was banned in animal feed in 1979, even though no residues in commercial meat were detectable.

Long before the ban, the effectiveness of DES, as well as concerns about its use, had stimulated research into other hormones as potential growth promoters. Since animals naturally produce estradiol, progesterone, and testosterone, these were ideal candidates. The stumbling block had been the cost of creating synthetic versions of these hormones, but once this problem had been overcome, they joined DES as feed additives and implants. By the time that DES was banned, the natural hormones, along with two synthetic compounds (zeranol and melengestrol acetate) gave farmers an ample choice of growth promoters.

Five years after DES had been removed from the market, Italian researchers published a paper in which they attributed an epidemic of breast enlargement in Italian school children during the late 1970s to baby food made with homogenized veal. Their evidence was estrogenic activity consistent with that of DES found in a third of baby-food jars that had been randomly collected. Hardly conclusive evidence, but enough to energize European consumer groups to lump all hormones together and mount an attack on their use in animals. In 1982 the Italian researchers reported that they had found no evidence of any DES in any baby food, and other scientists had suggested that inappropriate black market use of the substance probably accounted for the residues found earlier. Paying heed to consumers' concerns, European agricultural ministers asked a committee of scientists to investigate the issue. "There are no scientific grounds for a ban on either natural or synthetic growth promoters," was the conclusion. Still, a ban was implemented because "the ministers decided to pay more attention to political realities than to scientific facts," as stated by the European Community's agriculture commissioner. Those "political realities" may have also included the possibility that the ban would preclude the importing of American beef and give a boost to local producers.

University of Nottingham professor Eric Lamming, who had chaired the scientific advisory committee, was clearly disappointed by the decision. "I never thought scientific evidence would be disregarded in favor of misinformed consumer pressure," he grumbled. But do we really have scientific evidence for the safety of hormone use? No. Science can never guarantee safety, it can only demonstrate harm. It is always possible that someone somewhere may be adversely affected by the trace amounts of hormones in meat. But consider the following: an adult male produces about 136,000 nanograms of estrogen every day. Now compare this with the four nanograms found in a 170-gram (six-ounce) serving of beef from a treated animal, or the three nanograms in one not treated with hormones. Or with the 28,000 nanograms of estrogenic compounds in a tablespoon of soy oil. Consider also that an egg has forty-five times more estrogen than a quarter-pound hamburger. And that beer contains far more estrogenic compounds than meat, to say nothing of birth control pills or hormone replacement supplements. In light of such vast exposure to hormones, it is hard to imagine that the tiny amounts found in meat would be of any consequence. The improper use of hormones in animals by irresponsible producers is, of course, always a concern. There may also be a legitimate concern about hormones that end up in manure and eventually in natural water systems. Still, it makes more sense to worry about the saturated fat in meat, or its propensity to form carcinogenic compounds when broiled, grilled, or fried, than about its hormone content.

PCBs IN FISH

It's a pretty common scenario these days. Scientists publish a paper about finding some man-made pollutant in a consumer product and warn people about excessive exposure because the substance is known to cause cancer or reproductive problems when fed to rodents in high doses. The findings make front-page news and spokespeople for the industry in question complain bitterly that the risk has been exaggerated, while environmental groups hail the study as a breakthrough. Scientists with impeccable credentials wade into the debate on both sides, sometimes accusing each other of having vested interests. Different government regulatory agencies can't agree on what recommendations to make. The public is thoroughly confused. My office gets lots of e-mails and phone calls.

A recent scare was triggered by a paper in the prestigious journal *Science*, in which researchers report that farmed salmon are significantly more contaminated with organochlorine compounds such as PCBs, dioxins, toxaphene, and dieldrin than their wild counterparts. PCBs were once commonly used insulating fluids in electrical equipment; dioxins are by-products of some industrial processes; and toxaphene and dieldrin are insecticides. These chemicals are particularly persistent in the environment, and because they are fat soluble they accumulate in farmed fish that are fed fish meal and oil made from smaller fish. Similarly, when we eat contaminated fish, the organochlorides can build up in our fatty tissues. Everyone agrees that these compounds are capable of producing some pretty nasty health effects.

Let's use PCBs as an example and examine the cancer risk. There is no question that PCBs can cause the disease in animals, with the liver being the main organ affected. The human picture is less clear. Epidemiological studies have shown that workers with extensive exposure to PCBs in an industrial setting suffer a slightly elevated risk of cancer. Some investigators have also found a significant association between PCB concentrations in fatty tissue and non-Hodgkin's lymphoma. A couple of incidents in Japan and Taiwan in which people ingested rice oil accidentally contaminated with a high dose of PCBs also suggest an increased risk of liver cancer. Labelling PCBs as probable human carcinogens therefore seems justified. But that does not mean that eating fish, farmed or otherwise, raises the risk of cancer. As I've said several times before, our food supply contains numerous carcinogens, both natural and synthetic. Hydrazines in mushrooms, heterocyclic aromatic amines in cooked meat, aflatoxins in molds, and acrylamide in baked goods are all carcinogenic. But our diet also contains anticarcinogens in the form of various vitamins and polyphenols. When we eat, we consume hundreds of different chemicals, and the result of their interplay in our body is virtually impossible to predict. That's why the appropriate question to ask is not whether organochlorine contaminants in fish can cause cancer, but whether or not a diet high in fish does so. I am unaware of any study that shows a link between increased fish consumption and cancer. On the other hand, numerous studies point to just the opposite conclusion.

Swedish researchers have clearly shown that eating fatty fish, salmon in particular, can reduce the risk of prostate cancer by a third. Italian and Spanish scientists investigated the relation between frequency of fish consumption and cancer, and found that there was a consistent pattern of protection against the risk of digestive tract cancers, particularly of the colon, one of the leading causes of cancer mortality in developed countries. At the Cancer Center Hospital in Aichi, Japan, scientists looked at the diets of more than 4,000 healthy people and another 1,000 with lung cancer. Both men and women who ate large amounts of fresh fish were significantly less likely to develop lung cancer. This may explain why the Japanese, who smoke more than Westerners, have a lower rate of lung cancer. An extensive survey over ten years, involving more than 60,000 people of Chinese descent in Singapore, found that women who eat at least forty grams (1.4 ounces) of fish a day reduced their risk of breast cancer by 25 percent. There is sound theoretical justification for these observations. Prostaglandins are a class of chemicals in the body with a variety of hormonelike effects, some of which are linked to cancer. They are derived from arachidonic acid, which in turn forms from linoleic acid, a common omega-6 fat in the diet. Fish oils inhibit the cyclooxygenase-2 enzyme, which converts arachidonic acid to the problematic prostaglandin E2. Essentially then, cutting down on fish intake is likely to result in more, not less, cancer, irrespective of contaminants.

While the prospect of cancer instantly strikes fear into the heart, the fact is that strokes and heart disease kill more people. And there is overwhelming evidence linking fish consumption to protection from strokes and heart attacks. But why stop with cancer and heart disease? Recent evidence indicates that fish consumption offers protection from type 2 diabetes, and maybe even from Alzheimer's disease. In all cases, the beneficial chemicals are believed to be the omega-3 fats, of which salmon may be the richest source. Furthermore, salmon is less likely to be contaminated with mercury than other common fish.

What consumers should ask themselves is whether they should put more emphasis on the theoretical risks of organochlorides in fish or on the proven benefits of fish consumption. Although the answer should be obvious, the salmon study in *Science* is still an important one. It will undoubtedly encourage fish producers to take steps to reduce the organochloride residues in their product, something that is technically feasible. The use of feed made from canola and soy oil genetically modified to contain more omega-3 fats is an interesting possibility. Incidentally, canned salmon almost always comes from wild Alaska salmon, which are minimally contaminated with organochlorides. Most fish-oil supplements (the usual recommended dose is 1,000 milligrams a day) are also free of these compounds. I do, however, believe that the *Science* authors' argument that more than one meal of farmed salmon a month may hike the risk of cancer is totally unjustified. Since wild salmon is far more expensive, the warning about farmed salmon could have the effect of significantly reducing people's salmon consumption, thereby increasing the risk of illness. Pregnant women should stick to wild salmon, just to be ultrasafe.

TRANS FATS

You may not have heard of Tiburon, California, but in 2004 it became North America's first "trans fat–free city." New York followed in 2006, hoping to put a significant dent into the incidence of deaths from heart disease as it passed a law requiring restaurants to eliminate artificial trans fats. Health officials speculate that every year 500 New York deaths, more than the number of people killed in car accidents, can be prevented by removing trans fats from the diet. Consumers who want to avoid trans fats in processed foods can already do so since their presence must now be declared on food labels. This isn't enough for Pat Martin, a Canadian member of Parliament, who during a debate on labeling dramatically commented that "it is not okay to put poison in our food even if it is properly labelled." What then is this "poison" that is in our food supply and why is it there?

Trans fats entered our food supply as inadvertent by-products of hydrogenation, a process originally introduced as a health measure. To understand the chemical nuances involved, we need a little primer on fats. All fats are composed of a backbone of a three-carbon glycerol molecule to which long chains of carbon atoms, known as fatty acids, are attached. Each of these carbon atoms can bear a maximum of two hydrogen atoms, and when this is the case, the fatty acid is referred to as being "saturated" with hydrogen. If two of the carbons in the chain are attached to each other by a double bond, we use the term *monounsaturated*, because now there are two fewer hydrogen atoms than in a saturated fat, meaning the molecule is now *unsaturated* in terms of hydrogen. If more than one double bond is present, the molecule is termed *polyunsaturated*.

As a general rule, vegetable fats are either mono- or polyunsaturated (except for palm and coconut oils), while animal fats tend to be saturated. Saturated fats increase blood cholesterol and therefore have become nutritional pariahs. On the other hand, they are more suitable for baking and frying, because unlike unsaturated fats they do not break down when exposed to oxygen at high temperatures. Furthermore, saturated fats tend to be solids, meaning that they can be spread on bread more easily.

When the link between saturated fats and heart disease became evident, food producers, at the urging of health authorities, began to cut down on the use of saturated fats. But this was not just a simple matter of replacing these with the "healthier" polyunsaturated fats. The polyunsaturates did not yield the same texture and taste in foods, and they could not be used repeatedly in frying, which is crucial to the fast-food industry. Linolenic acid, present in virtually all vegetable fats, was particularly unstable to heat and prone to producing a rancid flavor on exposure to oxygen. So a compromise between saturated and unsaturated fats was needed and a process known as *partial hydrogenation* seemed to fit the bill.

Hydrogenation involves treating unsaturated fats with hydrogen gas under high pressure in the presence of a metallic catalyst such as nickel. Some of the double bonds react with the hydrogen, resulting in molecules that have fewer double bonds than polyunsaturated fats, but more than found in saturated fats. The newly created partially hydrogenated fats replaced beef tallow in frying and were also suitable for use in baked goods. Since they were solids, these fats also found extensive use in margarines, which were now portrayed as a healthier alternative to butter.

At the time, nobody suspected that this "healthier" alternative had a dark side. That's because nobody paid much attention to the fact that during the hydrogenation process some of the remaining double bonds were reconfigured from their naturally occurring *cis* form (hydrogen atoms on the same side of the double bond) to a *trans* arrangement. The effect was to straighten the carbon chains, which was initially perceived as beneficial because it allowed the chains to be packed more closely, solidifying the fat. And so it was that *trans fats* entered the marketplace. Soon they were everywhere. Crackers, pies, cookies, french fries, potato chips, breads, and margarines were full of them. Fine, everyone thought: trans fats belonged in the *unsaturated* category and were better for us than the saturated fats they replaced.

But then in the 1980s we began to hear some unsettling rumblings. Martijn Katan at the Agricultural University in Wageningen, the Netherlands, noted that while Scandinavians consumed more saturated fats than Americans, they had a lower incidence of coronary disease. Could this have something to do with American food producers' penchant for trans fats, he wondered? Dr. Katan decided to investigate. Volunteers were asked to follow a diet that featured monounsaturated fats, saturated fats, or trans fats. It was no surprise that people who consumed saturated fats had higher levels of LDL (the "bad cholesterol") and lower levels of HDL (the "good cholesterol"). But unexpectedly, the volunteers on the trans fat diet fared even worse than those on the saturated fat diet. The ratio of total cholesterol to HDL, an accepted measure of heart disease risk, rose 23 percent on the trans fat diet but only 13 percent on the saturated fat diet. True, the amount of trans fat eaten in this trial was more than the typical North American intake of 5 percent of total calories, but still, the point was made: trans fats increased the risk of heart disease.

The Nurses' Health Study, which has followed thousands of American nurses for over thirty years, corroborated the Scandinavian findings. Women who ate more cakes, cookies, white bread, and certain margarines, all major sources of trans fats, had a higher risk of heart disease. When blood samples were examined, researchers found that the amount of trans fat in red blood cells correlated significantly with the amount of trans fat consumed and was associated with increased levels of "bad" LDL cholesterol and decreased levels of "good" HDL cholesterol. They even managed to quantify the risk associated with trans fat consumption. Women with the highest trans-fatty acid content in red blood cells were three times more likely to develop heart disease than women with the lowest trans-fatty acid content. Other research has linked trans fats with type 2 diabetes, breast cancer, sudden cardiac death, asthma, and an increased risk of inflammation. Trans fats sure sound nasty. Indeed, a review of the scientific literature on trans fats published in the *New England Journal of Medicine* presents a pretty scary picture. It seems that cutting down on trans fats may not be enough; we may need to eliminate them completely from our diet. An analysis of four large trials involving some 140,000 subjects revealed that an increase of just 2 percent in caloric intake of trans fats was associated with a 23 percent higher risk of coronary heart disease. This means that even a couple of grams a day may be risky! The researchers make the startling prediction that as many as a quarter million coronary events a year could be avoided in North America by reducing trans fat consumption.

And now it seems that trans fats may affect our brain as well. At least that's the implication of Dr. Anne-Charlotte Granholm's research at the Medical University of South Carolina. Dr. Granholm trained rats to find a hidden platform in a water-filled maze. The animals were then put on a trans fat diet or a polyunsaturated fat diet and were asked to recall their training. The polyunsaturated rats swam right to the platform, whereas the trans fat rats floundered about. Just what happens on a molecular level isn't clear, but the theory is that somehow trans fats may cause inflammation that damages specific proteins involved in the transmission of information between nerve cells. And lest you think that the animals were overdosed on trans fats, such was not the case. The amount they consumed was typical of the North American diet. Dr. Granholm was disturbed enough by her results to swear off french fries and to rid her kitchen of processed foods that had trans fats.

Manufacturers are heeding the advice coming from researchers and are trying to reduce the trans fat content of their products. One way to do this is by resorting to oils that are low in polyunsaturated fats such as linolenic acid. As we have seen, it is the polyunsaturated fats that are unstable when heated and react with oxygen to produce off flavors. Corn and sunflower oils contain less than 1 percent linolenic acid but are more expensive than soy oil, which contains about 8 percent. Of course, the linolenic acid in soy oil can be hydrogenated, but then we have the problem of trans fats.

Recently an alternative approach has emerged. Low-linolenic-acid varieties of soybeans have been developed through traditional cross-breeding techniques to produce an oil that contains less than 3 percent linolenic acid and can therefore be used without hydrogenation. Given that the food industry uses the stunning amount of more than five billion pounds of frying oil a year, the potential market for low-linolenic acid soybean oil is huge. No wonder that farmers are rushing to plant the novel variety of soy. Other ways of eliminating trans fats also exist. A totally hydrogenated oil has no double bonds, hence no trans fats. Soy oil (or another polyunsaturated oil) can be totally hydrogenated, converting it to a solid waxy substance, which then, through a process known as *interesterification* can be reacted with a liquid polyunsaturated fat to produce a frying oil free of trans fat.

Food producers eager to get a step up on the competition are rushing to rid their products of trans fats. The Kellogg Company has already announced that it will replace its trans fat–laden oils in products like Pop Tarts and Cheez-Its with Vistive, a low-linolenic-acid oil produced by Monsanto. Predictably, this move has raised the ire of anti-genetic-modification activists who go into a frenzy any time Monsanto's name is mentioned. Actually, the trait for low-linolenic-acid content was introduced by traditional cross-breeding techniques, not through recombinant DNA technology. But the soybeans used to make the oil, like most soybeans grown in North America, also contain the trait for resistance to the herbicide glyphosate, so they do belong in the genetically modified category.

A more appropriate concern about low-linolenic-acid soybeans than whether they are genetically modified or not is whether they will make a significant enough impact on health. Let's face it, the foods burdened by their trans fat content are not the most nutritious foods to start with. Our consumption of potato chips, french fries, Pop Tarts, and Danish pastries should be limited in any case, no matter what kind of oil they are made with. Yes, technically speaking, if they are made with low-linolenic-acid oil, they are "better for you," but how much of a difference this makes to overall health is debatable. As far as snacks go, apples don't have trans fats. Nor do oranges. Or bananas. Or broccoli. Munch on those instead of doughnuts and you'll be healthier. And if Dr. Granholm's rats are any indicator, you'll be smarter too.

Having information about trans fats on food labels is a positive move. People wishing to avoid them can do so. Let's remember, though, that the numbers about lives saved by eliminating trans fats are based upon theoretical calculations, not hard evidence. In the last two decades the rate of heart disease in North America has declined significantly while trans fat consumption has stayed constant. Still, there is no downside to eliminating trans fats. And it can be done. In Denmark, foods containing more than 2 percent trans fats cannot be sold and the Danish food industry has not collapsed. But let's not jump to the conclusion that eliminating trans fats from cakes, doughnuts, or fries makes these foods "healthy." And let's not assume that eating a Danish for breakfast in Denmark is fine, but that it is "toxic" in North America. If you want a healthy breakfast, eat your oats, flax, and fruit. No worries about trans fats there!

To add a little more confusion to the trans fat story, it turns out that they are not all villains. There is no doubt that the ones that are a by-product of hydrogenation of unsaturated fats, the ones that show up in margarine and many baked goods, are decidedly unhealthy. But not all trans fats are manmade. Some occur in nature, and these *conjugated linoleic acids*, or CLAs, have quite different properties. They are mostly found in dairy products such as whole milk and cheddar cheese. The richest choice, believe it or not, is Cheez Whiz. Beef, lamb, and goat meat contain some CLAs as well. Bacteria in the guts of animals convert linoleic acid, a fatty acid found in animal feed, into CLA, which is stored in muscle and mammary tissue. We humans can't produce it ourselves, but research suggests that conjugated linoleic acids may be effective in fighting cancer, heart disease, type 2 diabetes, and weight gain. Rabbits fed a high-cholesterol diet are protected against heart disease if they also consume CLAs. This fat lowers their triglycerides (blood fats) and reduces their LDL. In rats, CLAs work as an insulin-sensitizer. Type 2 diabetics can't produce enough insulin, and CLAs may be a partial solution to their problem. In any case, CLAs lower triglycerides, which are often high in diabetics. Not all CLAs are created the same, only the ones referred to as *cis-9, trans-11* and *cis-10, trans-12* isomers are biologically active.

Perhaps the most alluring property that CLAs possess is their effect on controlling the body's muscle-to-fat ratio. In a three-month, placebo-controlled study, they significantly increased lean body mass in overweight patients. Over the twelve-week period, subjects taking 3.5 grams of CLAs a day experienced a 1.7 kilo reduction in pure fat. According to one of the world's leading experts in this field, Dr. Michael Pariza of the University of Wisconsin, the real potential of conjugated linoleic acids has to do with their ability to prevent weight gain (as fat) after weight has been lost. Pariza himself takes three to four grams daily.

The CLA connection to health is fascinating, but that's no reason to start gorging on meat and high-fat dairy products. The benefits that study subjects achieved came from doses far higher than the 100 milligrams found in an average diet. So, if further research does prove that CLAs are truly beneficial, supplements, not Cheez Whiz, will be the way to go.

BENZENE IN BEVERAGES

At one time benzene was used as an aftershave because of its sweet smell. It was even used to decaffeinate coffee. Clementine Churchill (Winston's wife) washed her hair with it. Oh my, how times change! Today we worry about a few parts per billion of benzene in our drinking water or in our soft drinks. Why? Because benzene is an established carcinogen and should be avoided. But benzene also happens to be a vital component of many products and processes essential to modern life. Traces of it are everywhere. Given that eliminating benzene from the environment is impossible, what we need is a reasonable risk analysis.

Not all appearances of benzene are due to human activity. It is one of the numerous compounds formed when organic matter decomposes, and therefore it can be found in petroleum. It also forms when organic matter burns, so volcanoes and forest fires produce benzene. So does burning coal. Michael Faraday, the brilliant English chemist and pioneer of electricity, first isolated benzene from "illuminating gas" in 1825. Back then, combustion of coal or peat produced the gaslight that illuminated homes and streets. The molecular structure of benzene, though, remained a mystery for some forty years. Chemists could not figure out how the six carbon atoms and six hydrogens that made up benzene were joined together. At least not until 1865, when German chemist August Kekulé had a dream in which a snake seized hold of its own tail. This vision led him to suggest a structure for benzene in which the six carbons were joined in a ring. Kekulé may have dreamt of a structure for benzene, but he could have hardly imagined the role this compound would play in building the industrialized world.

Today benzene is produced in huge amounts from petroleum feedstock. It serves as the raw material for making plastics such as nylon, polystyrene, and polycarbonate, as well as adhesives, detergents, dyes, insecticides, synthetic rubber, explosives, and drugs. Without a doubt, benzene makes our lives easier, but does it also make it shorter?

The first hint of potential health problems appeared when workers exposed to benzene vapors complained of dizziness, headaches, tremors, and even delirium, all symptoms of neurotoxicity. This led to the implementation of measures to reduce benzene exposure in the workplace, but lingering concerns about long-term exposure to small amounts remained. And it turned out that these concerns were justified: epidemiological studies eventually revealed a higher rate of leukemia in workers who inhaled benzene over a period of many years. The connection is not an overwhelming one; it is estimated that since 1928 when the association was first noted, there have been about 150 cases of leukemia worldwide that can be linked to occupational benzene exposure.

Most of us don't have to worry about occupational exposure. But what is the level of risk associated with benzene that shows up in our food and drink and in the air that we breathe? In theory, no amount of a carcinogen is safe, because a single molecular insult to DNA can lead to cancer. But, in practice, complete elimination of the numerous carcinogens to which we are exposed, both natural and synthetic, is impossible. In the case of benzene, most authorities have set a maximum allowable level in drinking water of five parts per billion. This does not mean that levels higher than this should prompt an immediate consultation with an undertaker. The standard of five parts per billion was set because it is achievable by municipal water treatment systems.

Now let's put some numbers into the benzene–cancer equation. We have two sources of information, human exposure data and animal feeding studies. When workers are exposed to less than 0.1

parts per million of benzene in the air, there is no evidence of increased risk of leukemia. Since we know the average human inhales about twenty cubic meters of air a day, this translates to an exposure of six milligrams a day. Animal feeding studies have confirmed that at such doses there is no increased cancer risk. So how much benzene are we exposed to? In 2006, we heard about a concern about benzene in soft drinks. At issue is the reaction of sodium benzoate, a preservative, with vitamin C, which is present in many beverages. It seems clear that in the presence of trace amounts of metals that catalyze the reaction, vitamin C produces free radicals that can convert benzoate into benzene. That's why some beverages have been found to contain as much as fifty parts per billion of benzene, ten times what is allowed in drinking water. But when we make the calculation, we find that a liter of such a drink contains fifty micrograms of benzene, which means that even at an impossible consumption rate of 120 liters a day, we would be below the amount that has no effect in occupational workers.

Of course, soft drinks are not our only exposure to benzene. When the US Food and Drug Administration carried out a survey of seventy foods over five years, benzene was found in every item except for American cheese and vanilla ice cream. A hamburger, for example, has four micrograms, but this is only one-tenth of the amount of benzene in the smoke inhaled from a cigarette. A banana can harbor up to twenty micrograms. Still, when all exposures are added up, we are well below the levels that have been linked with leukemia.

Is the risk zero? No. In some unlucky person a trace of benzene may start a cascade of events that leads to cancer. Therefore all efforts should be made to minimize exposure to carcinogens, particularly in the workplace, but traces of benzene in soft drinks are not a big deal. In any case, these can be eliminated by switching to preservatives other than sodium benzoate.

If you want to worry about something, worry about the lack of nutrition in soft drinks. Or about the benzene you're inhaling when you're pumping gas. That's about twenty micrograms. But even if you let the gas station attendant pump your gas, you'll be breathing in between twenty and thirty micrograms per hour just from the exhaust of the cars in front of you. In comparison, the average total daily intake from diet is about five micrograms. Such an analysis also holds true for the other volatile organic compounds (VOCs) of concern in our environment. When the FDA carried out its study of the food supply, it investigated the presence of over twenty other VOCs widely used as solvents, cleaning agents, degreasers, or intermediates in various chemical processes. Some could even enter the food supply as by-products of water chlorination or as migrants from plastics. Accordingly, chemical analyses were carried out for the likes of styrene, chloroform, carbon tetrachloride, and trichloroethylene, all of which are suspected carcinogens. All of these were found to be present at the parts per billion level, but like benzene, the amounts were well below toxic levels.

As an example, the minimum risk level for carbon tetrachloride has been established at 0.02 milligrams per kilogram per day. Some carbon tetrachloride can be detected in hot dogs at eleven parts per billion. A man weighing 70 kilograms (154 pounds) would have to eat 120 kilograms of hot dogs a day to reach this level. The bottom line is that volatile organic compounds are inhaled at much higher doses through cigarette smoke, gasoline fumes, or industrial emissions than they are found in foods and beverages.

TRANS-4-HYDROXYNONENAL IN FRIED FOODS

The Wiener schnitzel was so large it hung off the plate. Topped with a sprinkling of chopped parsley and lemon juice, it was an absolute treat. To this day, my mouth waters whenever I recall my first schnitzel experience. My aunt, who had arranged for us to come to Montreal after we had fled Hungary during the 1956 uprising, owned the Riviera, a European-style restaurant. It was there that I was introduced to the delights of a serving of veal, pounded almost paper thin, battered in flour, eggs, and breadcrumbs, and quickly fried to a golden brown. I just loved it. In fact, I still do. The Riviera is long gone, but I have learned to make a pretty acceptable schnitzel myself. But there is a difference. Science has entered the picture and my enjoyment is now tainted by nutritional concerns. Much as I hate to admit it, some pretty dark clouds hang over the frequent consumption of red meat and of fried foods.

The words *red meat* and *cancer* now appear in the same sentence in scientific literature with alarming frequency, and articles about the role of diet in cancer commonly conclude that many cases can be prevented by dietary modification. The suggested changes usually involve increasing fruit and vegetable consumption while curbing the intake of red meat and foods cooked at high temperatures. Take, for example, a huge European study that enrolled almost 500,000 healthy men and women in the 1990s and followed their health status. After about five years, some 1,300 cases of colorectal cancer had been detected. The lifestyles of these patients were then compared with those free of the disease. The major finding was that bowel cancer was associated with an intake of red meats and processed meats. Quantitatively, people who ate more than 160 grams (5.6 ounces) of red or processed meat a day were 35 percent more likely to develop bowel cancer than those who ate less than 20 grams a day. Chicken was not implicated and eating fish was associated with a lower risk of bowel cancer.

Exactly what the problem is with red and processed meats is hard to say, but it's a good bet that heterocyclic amines (HCAs) are involved. Heating food unleashes a host of chemical changes, some of which, such as destroying bacteria, softening muscle fibers, and developing flavor are desirable, while others are not. High temperatures allow compounds such as creatinine in meat to engage in reactions that form heterocyclic amines, which are recognized carcinogens. The higher the temperature and the longer the cooking time, the more HCAs form. And these compounds have been implicated in more than bowel cancer. Red-meat consumption is associated with prostate, stomach, and pancreatic cancer, and researchers have also found that women who routinely eat very well-done meat face a fivefold increase in breast cancer risk when compared with women who eat their meat rare or medium. Why chicken and fish are less risky isn't clear but it may have to do with shorter cooking times. In any case, this is a welcome observation because chicken and (especially) fish are also deemed to be more heart healthy than red meat—as long as they are not fried. Harvard Medical School researchers who examined the heart function of some 5,000 seniors found that those who ate broiled or baked fish frequently had lower heart rates, lower blood pressure, and better blood flow to the heart, while those who regularly ate fried fish or fast-food fish sandwiches showed a greater incidence of hardening of the coronary arteries and other heart problems. The likely culprit here is the fat used for frying.

I don't know what the cooks in the Riviera used to fry my Wiener schnitzel back in the 1950s, but I suspect it was some sort of animal fat. Today, we are more knowledgeable about nutrition and would lean toward using polyunsaturated fats, hopefully devoid of trans fats. But that doesn't mean we have licked the problems associated with fried foods. In fact, a new suspicious compound is emerging, trans-4-hydroxy-2-nonenal, or HNE. How's that for a mouthful?

Chances are you haven't heard of HNE, but it is causing somewhat of a ruckus in the scientific community. HNE forms when polyunsaturated fats (those containing several carbon-carbon double bonds) react with oxygen. Such fats are present in cell membranes and can give rise to HNE, which then can travel through the bloodstream. The bad news is that HNE has been linked with cardiovascular disease, Parkinson's, Alzheimer's, liver and kidney ailments, and even cancer. And here is what we really don't want to hear: HNE forms when polyunsaturated oils, particularly those containing linoleic acid (corn, soy, canola), are heated, especially if heated repeatedly. Those golden fries in restaurants may be laden with HNE!

Now for the good news. Monounsaturated fats like peanut oil or olive oil are far less prone to such contamination. Alas, these are not commonly used in restaurants, so limiting fried foods when eating out is really important. But I haven't given up on making my Wiener schnitzel at home; I just do it less often and fry the veal in olive oil. There is more to life than worrying about every morsel of food we put into our mouths.

SUBSTANCES LEACHING FROM PLASTICS

We wrap our leftovers in plastic film. We use plastic bags for sandwiches, fruits, and vegetables. Our meat often comes wrapped in plastic. We buy beverages in plastic bottles, we often use plastic cutlery, drink from plastic glasses, and microwave in plastic dishes. As a consequence we ingest dozens of substances that leach out from plastics into our foods and beverages: plasticizers that make plastics soft and pliable; stabilizers and catalysts used in linking small molecules (monomers) into the long chains (polymers) that characterize plastics; residual monomers and some polymer decomposition products. All of these can end up in our bodies. Does this matter? According to the folks who circulate scary e-mails, the answer is yes. The allegations are that cancer-causing substances such as plasticizers and dioxin leach out of plastics and that "Saran Wrap placed over foods as they are nuked with high heat actually drips poisonous toxins into the food."

The e-mail begins with the captivating saga of Claire Nelson, an inquisitive high-school student in Arkansas who learned that a plasticizer called di(ethylhexyl)adipate (DEHA) is found in plastic wrap and that the US Food and Drug Administration had never studied whether this "carcinogen" migrates into food during microwave cooking. With the help of a professional scientist, she devised an experiment in which she cooked up a mix of plastic wrap and olive oil and found that DEHA migrated into the oil at levels far higher than the FDA standard of 0.05 parts per billion. Nelson eventually received the American Chemical Society's top prize for students and her story charmed many a reporter. They were keen to portray her as the people's champion who had uncovered yet another attack on the public's health by an uncaring industry aided and abetted by an incompetent FDA.

Claire Nelson is real and did win a prize for her work. But the prize was for her systematic investigation of a possible problem, not for her role in unmasking a cancer threat. Indeed, there was no threat to be unmasked, since the migration of DEHA into food had been studied before. The notion that Nelson was the first to think of this idea is romanticized folklore. After all, the FDA already had a standard in place for acceptable levels of DEHA, the standard that her results exceeded. This really came as no great surprise. Heating plastic wrap immersed in oil for extended periods to study migration of plasticizer into the oil is hardly a realistic situation. It is akin to trying to evaluate the risks of city driving by studying Formula 1 racing.

In any case, is the plasticizer as dangerous as portrayed? These chemicals are commonly added to plastics to make them soft and pliable. Shower curtains are a typical example. Plasticizers are also used to improve "cling" in certain food wraps. Concerns have arisen because of the possibility that some, particularly di(ethylhexyl)phthalate (DEHP), may have estrogen-like properties, which in theory can be linked to certain cancers. But DEHA, the plasticizer that is used in polyvinyl chloride (PVC) wrap, does not fall into this category. Both the European Union and the Environmental Protection Agency in the United States have now classified it as "not a suspected carcinogen." This is the plasticizer that Claire Nelson studied.

Only PVC wraps are plasticized with DEHA. While these are commonly used in commercial food packaging, they are not the wraps that consumers are likely to purchase and use in their microwave ovens. Glad Wrap, for example, is made of low-density polyethylene (LDPE) and has no phthalates at all. The same goes for Saran Wrap. Saran Wrap used to be made of polyvinylidene chloride, which had excellent barrier and cling properties but was replaced by LDPE in 2004 as the manufacturer switched to reduce the environmental footprint of chlorinated compounds. LDPE itself is not clingy enough, but its ability to stick is improved by the incorporation of other polymers, such as polyisobutene or linear low-density polyethylene. Neither of these presents any concern. Even when Saran Wrap was made of polyvinylidene chloride, the plasticizer used was acetyltributyl citrate, so there never was a "phthalate issue" with the product. It is hard to imagine, then, what "poisonous toxin" (are there any nonpoisonous toxins?) could have "dripped into food" from Saran Wrap. It does stand to reason that any plastic wrap should be kept out of direct contact with food in a microwave for the simple reason that food, particularly if high in sugar or fat, can get very hot and melt the plastic. Eating melted plastic may not be dangerous but it is unpalatable.

But what about the accusation that heating food in plastic containers in the microwave will cause the transfer of carcinogenic dioxins into the food? Certainly, dioxins are carcinogens and we have to make every effort to avoid them. For plastics to release dioxins, however, two conditions have to be met. They must contain chlorine and they have to be heated to incineration temperatures. The containers consumers use at home (Tupperware, GladWare, Rubbermaid) are made of polyethylene or polypropylene and cannot give rise to dioxins, for the simple fact that they do not contain any chlorine. Nor do the containers in which you bring home those delicacies from the deli counter; these are also usually made of polypropylene. As a general rule, such containers, including old margarine tubs, should not be used in the microwave, not because of any dioxin issue, but because they may soften or melt.

The only kind of common container that in theory could generate dioxins is one made of polyvinyl chloride (PVC). While PVC is used extensively in cleaning products and cosmetic packaging, it is not used to make microwavable food containers. Even if it were, temperatures in the microwave are not nearly high enough to break the plastic down and release dioxin.

Despite any alarming e-mails, there is no scientific basis for concern about using plastics in the microwave. There is reason, however, to be concerned about the ease with which unreliable information spreads via the Internet and about the unnecessary anxiety it creates. Another excellent example of this is the fear of perfluorochemicals, which are used in some packaging materials and in making Teflon cookware. Supposedly, these too are a source of carcinogens.

Consumers may be quite willing to put up with messy hands when they eat popcorn, but they certainly don't want to see an oil-stained package on the shelf. And that's where perfluorochemicals come in. Added to the packaging material, they impart grease-resistant properties. But unfortunately they also have a tendency to migrate into the oily goo that is added to the popcorn to simulate butter. There is an indication that such package coatings may be a source of perfluorooctanoic acid (PFOA), a compound found in the blood of virtually all North Americans, and PFOA is suspected of being a carcinogen.

Now before anyone starts to organize street demonstrations to ban microwave popcorn, a few thoughts on carcinogenicity are in order. By definition, a carcinogen is a substance capable of triggering cancer in people or animals. So far, some sixty substances have been classified as human carcinogens. These include asbestos, alcohol, certain arsenic compounds, benzene, tobacco smoke, soot, estrogen, mustard gas, radon, ultraviolet light, tamoxifen, vinyl chloride, and wood dust. Human epidemiological studies have clearly shown that exposure to these substances is linked with cancer. Furthermore, there are reasonable molecular mechanisms to explain how these chemicals can cause the disease. Dosage is important; you don't get cancer from smoking one cigarette.

Aside from established human carcinogens, there is a large number of substances known to be animal carcinogens, based on feeding studies. In most cases the dose to which the animals are exposed is so large that it is difficult to establish human relevance. Consider furfural, a compound used in some plastic manufacture, but one that also occurs naturally in grains, sweet potatoes, and even apples. There is no doubt that it is a carcinogen. Feed it to rodents at a dose of 200 milligrams per kilogram of body weight and it will cause cancer. Since bread is made from grains, it will contain furfural. By referring selectively to the scientific literature, one could then argue that bread can cause cancer. Panic in the pantry would ensue if a crucial little detail were left out, namely that a person would have to consume roughly 6,000 loaves of bread a day to approach the amount of furfural that causes cancer in rodents. Let's point out yet again that there are numerous other substances, both natural and synthetic, that can rightfully be labeled as animal carcinogens. Caffeic acid in coffee, acrylamide in french fries, safrole in black pepper, certain pesticides, PCBs, dioxins, and some fluorinated compounds fall into this category. But that does not mean that pepper or coffee causes cancer. In fact, we have good evidence that they don't. The carcinogens are there all right, but not in sufficiently high doses.

Now let's return to the PFOA issue. Thanks to phenomenal advances in analytical chemistry, we know that this chemical is present in most people's blood at a level of roughly five parts per billion. One part per billion is one second in thirty-two years, or one square of toilet tissue in a roll stretching from New York to London. Obviously, we don't have much PFOA in us, but why do we have any at all? Where is it coming from? Accusing fingers have been pointed at Teflon producers. The *emulsion polymerization* process by which this plastic is manufactured requires oily substances to be mixed with water. This is a job for chemicals called surfactants, and PFOA fits the bill perfectly. The surfactant is not present in the finished product, so Teflon pots and pans do not release it, at least not under ordinary cooking temperatures. At temperatures above 6,332°F (3,500°C), traces of PFOA may form if some Teflon decomposes, but this certainly cannot account for the PFOA that shows up in the environment.

Truth be told, DuPont, a major Teflon producer, until recently has been less than fastidious about containment of PFOA and has contaminated the water supply around its Parkersburg, West Virginia plant. This situation led to allegations of increased cancer rates in the community and a class action lawsuit that the company settled for over $300 million. DuPont officials did not admit any guilt and pointed out that the cancer studies did not control for possible causes other than PFOA. More recently the company was penalized $10.25 million by the Environmental Protection Agency for not having reported some toxicological studies it had carried out, one of which showed that PFOA was found in the umbilical-cord blood of a baby born to a woman working in the Teflon-producing plant. The fine was for not having reported the data, not for endangerment.

The release of PFOA from the plant, though, does not explain the widespread distribution of this chemical. Nevertheless, the EPA has asked manufacturers to reduce PFOA emissions by 95 percent by 2010 and to stop emitting it totally by 2015. DuPont has already announced that it will meet this goal even before the proposed deadline. In any case, eliminating PFOA from Teflon production will not eliminate the problem of the chemical showing up in blood, because this is not its major source. A more likely scenario, effectively demonstrated by University of Toronto chemist Scott Mabury, is that short-chain fluorochemicals, or *fluorotelomers*, which are widely used in food packaging, coatings, paints, fire-fighting foams, inks, adhesives, and waxes can break down in the environment, or in the human body, to release PFOA. Chemists will have to find alternatives for these substances.

What might happen if we do not reduce PFOA in the environment? It is a persistent chemical, that much is for sure. Researchers at Johns Hopkins University found it to be present in the umbilical cord of virtually every baby born at the university hospital. But is it causing any harm? So far, there is little evidence for this. Studies of DuPont workers who have been exposed to amounts in orders of magnitude greater than the public normally experiences have not revealed any increase in cancer rates, although there is a suggestion of elevated cholesterol levels. Indeed, the EPA's own risk assessment suggesting that PFOA is a possible carcinogen is based on very weak data. The rat studies are equivocal, and the report clearly states that "the mode of action by which PFOA may cause tumors in rats is unlikely to occur in humans."

Headlines such as "Teflon Chemical Causes Cancer" or "Dangers Lurk in Teflon Pans" are sensationalist misrepresentations of the facts. There is no evidence that at five parts per billion in the blood PFOA does any harm. Based on what we know so far, there is probably more reason to worry about the saturated fat in the microwave popcorn than about the fluorotelomers in the packaging. And you can always make old-fashioned popcorn. Just take a pan, heat a little oil, and add the kernels. And if you don't want the popcorn to burn and produce carcinogens, use a Teflon pot!

THE BISPHENOL A ISSUE

What was it that Alexander Pope said? "A little learning is a dangerous thing / Drink deep, or taste not the Pierian spring / There shallow thoughts intoxicate the brain / And drinking largely sobers us again." But the poet never had to think about drinking that spring water out of a polycarbonate bottle, did he? Today, such thoughts cannot be avoided. Bisphenol A (BPA), a chemical that can leach out of polycarbonate bottles or out of the resins that line cans, is clearly the "toxin du jour." Environmental organizations label it as a clear threat to our health, while the plastics industry maintains that the scare is exaggerated. So who is right? Neither side. BPA is not a major threat to health but neither can its possible effects be dismissed out of hand.

The oft-repeated tenet of toxicology is that only the dose makes the poison. We know that two aspirin tablets are safe, but swallowing a whole bottle of pills can be lethal. A blood sugar reading of 155 mg/dL (4 mmol/L) causes no concern, but 387 mg/dL (10 mmol/L) raises a red flag. However, in the complex world of toxicology, the interpretation of numbers is rarely that simple, with some researchers even challenging the notion that risk is proportional to increasing dose. Very tiny amounts, they claim, may have different consequences than larger doses, especially when it comes to chemicals with estrogen-like effects. Welcome to the world of *environmental endocrine disruptors*. Welcome to the world of bisphenol A (BPA).

BPA is the chemical that has people reaching for the panic button. And hoping that the button isn't made of polycarbonate (PC), a plastic that can release trace amounts of bisphenol A into the environment. Polycarbonate plastics are ubiquitous. Compact discs, DVDs, hockey helmets, eyeglasses, water pipes, automobile headlights, bullet-proof shields, dental sealants, cell phones, and laptops all benefit from the virtually unbreakable properties of polycarbonates. Bisphenol A is one of the components used to make polycarbonates as well as epoxy resins and can indeed be released in trace amounts from these materials over time. Refillable water bottles, baby bottles, and food-processor bowls made of polycarbonate can all leach BPA into their contents as can protective linings inside food cans. In addition, bisphenol A is used as an antioxidant additive in polyvinyl chloride (PVC) plastics such as garden hoses and in industrial food wraps.

Environmental estrogens raise eyebrows because they can mimic the activity of the body's own estrogen. That synthetic chemicals can behave in this fashion was first shown in the 1930s by E. C. Dodds and Wilfrid Lawson at Middlesex Hospital in London through the examination of vaginal smears taken from rats after removal of their ovaries. As expected, the cells were devoid of the changes normally triggered by estrogen released from the ovaries during estrus, the period when the female is receptive to the advances of a male. But injection of the rats with certain synthetic compounds resulted in the detection of a positive estrus response in the vaginal cells, indicating estrogenic activity. One of the compounds that Dodds and Lawson found to trigger such activity was bisphenol A, the substance that now appears in the blood and urine of virtually everyone living in the Western world. But unfortunately our ability to detect trace amounts of such chemicals in our bodies has surpassed our ability to interpret what the numbers mean. Hence the debate about the risks of BPA exposure.

And there certainly is debate! Producers of polycarbonate plastics maintain that the amounts to which we are exposed are inconsequential, while some researchers claim that BPA even at levels of parts per trillion, as found in our bodies, is potentially harmful. Accusations, sometimes with venomous overtones, fly back and forth. Supporters of the "bisphenol A is a health hazard" side accuse industry of distorting the numbers, while industry spokesmen are quick to drill holes in the research that damns BPA. It is quickly broken down in the body, they point out, and it is not BPA, but one of its breakdown products that is actually measured in the urine. This compound, they maintain, does not have estrogenic properties and is not an indication of the body load of BPA.

Furthermore, they argue, a human is not a giant rat. Indeed, studies have shown that rodents employ a different detoxification mechanism than humans, and do have more circulating BPA after exposure than we would have. Then there is the issue of type of exposure. Many of the rodent studies have used injected or implanted BPA, which is a different type of exposure than ingestion. But the BPA critics fight back. Injecting BPA into pregnant rodents is an appropriate way to study effects on the fetus, they retort, and they also point to some studies that show ingestion and injection lead to similar blood levels of BPA. And as far as humans go, they say, while indeed detoxification reactions do swing into action, these are much less efficient in children and babies, who are therefore at greater risk. And so it goes.

What, then, do we make of the battle of these scientists? Scrutiny of the findings by an independent panel of experts would seem to be a logical step. And that step has been taken, both in Europe, where regulations about toxic exposure are far more stringent, and in the United States. The European Food Safety Authority (EFSA) is made up of experts who provide independent advice on risk management. Based on the maximum dose that produces no observable toxic effects in animals (NOAEL), and building in a safety factor of 100, the EFSA in 2006 concluded that the Tolerable Daily Intake (TDI) for humans is fifty micrograms of BPA per kilogram of body weight per day. The panel then went on to calculate exposure for all age groups based on data from leaching experiments, and found that in all cases exposure was less than 30 percent of the TDI, which of course already has a huge safety factor built in. However, there is a fly in the ointment here.

The NOAEL does not take into account the subtle effects that may be noted in the offspring of animals treated with BPA. In the United States, the National Toxicology Program's Expert Panel addressed this issue in 2007 and expressed "some concern that exposure to Bisphenol A in utero, as well as in infants and children, causes neural and behavioral effects" but passed off all other concerns as negligible. This did not sit well with the Environmental Working Group, a high-profile watchdog organization in the United States that has accused the panel of "having been unduly influenced by industry and having issued a report that failed to meet the most basic scientific standards." With all this mudslinging, and with both sides spouting numbers that seem to back their position, what is the poor consumer to do? Should we be terrified of our water bottles and canned foods?

The current media focus on BPA was stimulated by a couple of studies that measured the amount of this chemical that leached into water stored in polycarbonate bottles. To gain some perspective, let's consider baby bottles and "worst-case scenarios," since if BPA presents a risk, it is expected to be most significant during early development. In the most recent study, the maximum amount of BPA that leached into water from a polycarbonate bottle was eight nanograms per milliliter. Let's assume a baby were to drink a liter of this water. One milligram is a million nanograms, so the total intake would be 0.008 milligrams. If the baby weighs 5 kilograms (11 pounds), we have an intake of 0.0016 milligrams per kilogram of body weight. This is about one-thirtieth the TDI and one three-thousandth the *no observable adverse effect level*, in test animals.

That's theory. What about measuring how much BPA we are actually exposed to? That's been done. The Centers for Disease Control in the United States sampled urine from over 2,000 people aged six to eighty-five and found an average of about 2.7 nanograms per milliliter. Since bisphenol A does not accumulate in the body, the urinary output can be used to estimate the amount taken in through food and water. This calculates to fifty nanograms per kilogram of body weight. And how does that compare to the TDI? It is 1,000 times less! And one hundred thousand times less than the highest dose that causes no effect in test animals! Some researchers argue that the calculation of oral intake based on urinary output is flawed, and that if one goes by studies in animals the value should be at least 100 times greater. Even if we were to accept this highly contentious argument, we are still looking at an intake that is a thousand times less than the highest dose that causes no effect in test animals. So there seems to be a pretty significant safety factor here, even if one wants to argue about the exact value of the NOAEL.

But there is an emerging issue here, one that calls into question the way the NOAEL is determined. In fact it calls into question the very foundations of toxicology, namely that "the dose makes the poison." The relatively novel concept of *hormesis* contends that at very low concentrations substances may exhibit unpredictable behavior. In other words, one cannot assume that a substance that shows no effect at a certain concentration will show no effect at lower concentrations. Perhaps an analogy may be appropriate here. Imagine that you have a large crowd of people trying to squeeze through a single door. With all the pushing and shoving, it may well be that nobody gets through. But if you have only a few people around, they can easily get through the door. And so it may be with hormones. At a low dose certain types of cellular activity may be triggered that may not occur at higher doses. Curiously, this effect may be detrimental or beneficial, as has been noted with BPA.

This brings us back to the problem that the NOAEL may not be an appropriate way to evaluate hormone-like effects. For that, we may have to turn to the field of epigenetics, which focuses on the ability of environmental chemicals to turn genes on or off in the fetus, possibly sensitizing the body to disease in adulthood. For example, when agouti mice are treated with doses of BPA that are five times less than the

NOAEL, their offspring face a greater risk of obesity, diabetes, and cancer even though their genetic makeup has not been altered. However, when the mice are also fed folic acid or genistein, an estrogenic compound found in soy, the effect of the BPA is negated.

There is also some evidence that young male rats, exposed to levels of bisphenol A comparable to that found in human blood, are more likely to develop precancerous prostate lesions as they grow older. Furthermore, female rats exposed to bisphenol A during their fetal stage are more likely to develop mammary cancers as adults. And researchers have also found that normal cells removed from patients during breast cancer surgery are more likely to become malignant when cultured in the laboratory in a solution containing very low doses of bisphenol A. There is also some preliminary evidence suggesting an association between higher blood levels of bisphenol A and a greater likelihood of miscarriage and polycystic ovarian syndrome. On the other hand, some researchers have found that at very low doses BPA provides a reproducibly clear protection to neurons against model neurotoxins such as beta amyloid and glutamate, which have been implicated in diseases such as Alzheimer's.

Each of these studies can be, and has been, subjected to valid criticism. Associations cannot prove cause and effect, and culturing cells in the lab is quite different from cellular activity in the body, where our cells are exposed to not only one specific chemical but to hundreds, both natural and synthetic. Many of these may have hormonal effects that may either exacerbate or neutralize those of bisphenol A. Also, in animal and cell culture studies, exposure is continuous, whereas this may not be the case with consumers. Since bisphenol A is quickly metabolized and excreted by humans, the risk of sporadic exposure may be quite different from that predicted by animal models based on continuous exposure.

There is also the possibility that we are more sensitive to hormone-disrupting chemicals than rodents. But if that is the case, we have a lot more to worry about than just BPA. Remember the joke about the drunk who was walking back and forth below a street lamp? What did you lose? he was asked. My keys, came the reply. Did you drop them here? No, he answered, but this is the only place where there is light! Right now, the light is being cast on bisphenol A, while numerous hormonelike substances lurk in the darkness.

Take lavender-scented soaps and lotions, for example. These have been linked with breast growth in young boys. As it turns out, lavender oil activates estrogen-regulating genes in human breast cells. Alfalfa-sprout extracts display increased breast cancer cell–proliferation effects above levels seen with estradiol, an estrogen. Soybeans contain natural estrogenic compounds, and so does milk. Milk represents a far greater estrogenic exposure than we experience from BPA. Our average daily intake of estrogens through milk is about 370 nanograms, which is roughly what would be found in 50 milliliters of water from a polycarbonate bottle. Nobody is suggesting the banning of milk even though it contains a good dose of estrogenic compounds. And neither should they.

This does not mean that we should be cavalier about hormone-like substances in the environment. Even though there is no evidence that at the levels encountered BPA presents a risk to humans, we can't rule out the possibility that babies may not excrete BPA as efficiently as adults, or that the chemical may have a synergistic effect when combined with other endocrine-disrupting substances. Keep in mind, though, that we are exposed to thousands of natural and synthetic compounds every day in our food, water, air, cosmetics, cleaning agents, and drugs, many of which have hormonelike effects and, if scrutinized with the same vigor as bisphenol A, would raise similar concerns. Health is an extremely complex business, with genetics, nutrition, and myriad environmental factors playing a role. Focusing on one specific chemical presents an unrealistic picture and creates an aura of importance that is out of proportion to the actual risk. The accumulating scientific evidence justifies a careful look at how bisphenol A enters the environment, but certainly does not justify the current wave of hysteria.

So how can we distill all this down to some practical advice? A polycarbonate bottle, recognizable by the number "7" in the recycling logo (although not all #7 is polycarbonate), is not the best choice for feeding babies, and the same goes for liquid formulas in cans. It's probably prudent, if for no other reason than emotional comfort, for someone who has had a bout with a hormone-sensitive cancer, to not overindulge in canned foods or beverages from polycarbonate bottles. For microwave cooking, glass or ceramic is best, but as far as plastics go, polyester (#1), polyethylene (#2), and polypropylene (#5) dishes have no BPA issue.

Finally, how about a little perspective? We talk about banning a substance that in theory may have some adverse effect, while we allow the sale of cigarettes, which are known to kill millions annually. And drunk drivers do their share as well. It seems we do need a sobering drink from that Pierian spring.

DIOXINS

"I wish I knew." That's the answer I had to give when asked if dioxins in meat or milk cause cancer. And this is the only scientifically legitimate response to such a question. There are some who contend that dioxins are such potent carcinogens that no amount in the diet is acceptable, while others claim that the trace amounts to which we are exposed are of no consequence. To get a handle on the situation, we turn to toxicology, the science that investigates the effects of chemicals on our health. Toxicology may not be able to provide absolute answers, but it does allow judgments to be made based on animal tests, knowledge of biochemical pathways, molecular structure, and human epidemiological data.

Toxicologists are very adept at predicting the acute effects of chemicals. We know that taking roughly 100 aspirin tablets at a time is likely to result in death. The lethal amounts of arsenic, cyanide, or strychnine are well established. But toxicology is on a much weaker footing when it comes to chronic effects; that is, long-term exposure to amounts way below those that produce acute catastrophes. Chronic effects cannot be predicted based on observations of acute toxicity. Vitamin D, for example, is acutely toxic, but in small daily doses contributes to good health. The caffeine content of 100 cups of coffee would kill an adult, but a cup of coffee a day is not toxic. It is well known that a single large exposure to a chemical can trigger a different biochemical response than is triggered by long-term exposure to small amounts. Acute exposure to chloroform, for example, causes dizziness followed by sedation. On the other hand, chronic exposure to small amounts of chloroform can cause liver damage. There is no doubt that a large exposure to dioxin causes chloracne, a disturbing skin condition. But that tells us nothing about whether or not trace amounts in the diet can be linked to cancer.

Why does the question arise in the first place? Because feeding dioxins to test animals in high doses can certainly cause cancer. Many scientists, though, express concern about the meaning of such studies. The supposition is that if a large dose produces cancers in test animals, the same cancers will be produced in proportion to smaller doses, no matter how small these are. Actually, we know that our bodies generate various enzymes that can repair the type of damage to DNA molecules that would initiate cancer. After all, we are exposed to potential carcinogens all the time, both natural and synthetic. Ultraviolet rays from the sun, benzopyrene in a charcoal-broiled steak, and alcohol in wine are all established carcinogens but our body appears capable of handling small doses. It is very likely that for carcinogens, as for other toxins, there is a "threshold effect," above which the body's protective chemistry is overwhelmed, but below which there are no concerns.

The standard animal tests for toxicity are based on the *maximum tolerated dose* (MTD), sometimes referred to as the *no observed adverse effect level* (NOAEL). This is the maximum amount of a chemical that can be given to animals without triggering any adverse consequence. If this amount is exceeded, animals get sick. If they develop cancer, the chemical in question is termed a *carcinogen*. Safe human exposure levels are then determined based on the maximum amounts that showed no effect in animals.

Actual human exposure to a chemical is usually estimated from blood tests or from the amount of the chemical found in the food supply. In many cases, the MTD is far in excess of possible human exposure. For example, if rats develop tumors when exposed to a chemical at a dose that is 101,000 times greater than what a human can ever encounter, but not at a dose that is just 100,000 times greater, it is still categorized as a carcinogen. A more reasonable approach might be to first determine the maximum human exposure, build in a multiplier safety factor of perhaps 100, and test this dose in animals. A great deal of undue worry about theoretical carcinogens that may have no practical significance could be prevented.

There are other concerns with animal tests. A human is not a giant rat. There are biochemical differences. Eye damage from large doses of methanol does not occur in rats, but does in humans and other primates. Nitrobenzene is far more toxic in man, dogs, and cats than in monkeys, rabbits, or rats. It takes 5,000 times as much dioxin to kill a hamster as a guinea pig. But what about cancer? Dioxins can certainly cause cancer in some animals at lower doses than other carcinogens. In rats, liver tumors form at a daily intake of ten nanograms per kilogram of body weight, but there is no effect at one nanogram per kilogram. The average human exposure is about 0.002 nanograms per kilogram, which is 0.2 percent of the highest dose that causes no effect in animals. Still, given that dioxins do occur in the environment, people are legitimately concerned.

Dioxins, of which there are some seventeen varieties with differing toxicities, are unintentional by-products of combustion and some industrial processes. They deposit from the atmosphere onto soil and plants, and can find their way into our bodies as we dine on crops or on animals that have eaten crops. Do these tiny amounts matter? We can get an idea from humans who have been exposed to far larger amounts. Numerous studies have investigated Vietnam veterans who were exposed to dioxins as contaminants in the notorious defoliant Agent Orange, workers in the herbicide industry, victims of a massive accidental release of dioxins from a chemical plant in Seveso, Italy, in 1976, and people living in the vicinity of incineration facilities. Some of these studies found slight increases in some cancers, some found no relationship to dioxins, and some even claimed to find a lower incidence. As far as diet goes, there is some evidence that dioxins, at a low dose promote cancer, but only when fed to animals after other carcinogens such as aflatoxins, found in molds, have been ingested. When fed before other carcinogens, dioxins result in lower cancer rates. And that's as much as toxicology can currently tell us.

PART FOUR

TOUGH TO SWALLOW

THE MIRACLE OF GOJI JUICE?

How does one become the "world's leading nutritionist"? Do you win a competition? Do other nutritionists get together and take a vote? Do you publish the most research papers? Or do you get the title bestowed upon you by some publicity firm engaged to sell your books and products? The latter seems to be the case for Earl Mindell, pharmacist, "master herbalist," and "doctor of nutrition." Mindell is an industry. He lectures, writes books, appears on television and radio, and, above all, makes amazing nutritional discoveries. Like goji juice, the wonder product that makes people "look and feel twenty years younger."

For someone who earned a bachelor's degree in pharmacy from the University of North Dakota, Mindell makes some curious statements. In one of his pamphlets, he talks about reversing the aging process by eating foods, such as sardines, that are high in DNA and RNA. The truth is that these nucleic acids are completely digested by our bodies and never reach cells to do any good. He has also promoted oral supplements of an "anti-aging" enzyme, superoxide dismutase (SOD). Not only is there no evidence for the supposed benefits of SOD, it would not survive the digestive process. Okay, maybe this isn't exactly pharmacy, so Mindell can be excused. But he also has a PhD in nutrition! And certainly a nutritionist would know about the chemistry of nucleic acids and enzymes. Mindell's so-called PhD comes from Pacific Western University, which has no classrooms, offers no lectures or labs, but grants degrees "without attending class because PWU recognizes the inherent value of prior education, training and work experience and knowledge accrued through past experience."

I guess we can then understand if "Dr." Mindell is a little out of touch with mainstream nutrition, which is based on laboratory research, epidemiological studies, and placebo-controlled trials. And why he may not appreciate that thousands of legitimate nutritional researchers working around the world have failed to come up with "miracles." Miracles are hard to come by in science—yet Mindell has had a career of "discovering" them. There was his *Soy Miracle* book. Then there was *Amazing Apple Cider Vinegar.* In his epic *Russian Energy Secrets,* Mindell describes how you can fight cancer, heart disease, and liver problems by using sixteen magical herbs. And now Mindell has made his "most important health discovery ever": Himalayan goji juice.

Mindell tells us this Asian remedy has been used for "countless generations" to solve all sorts of health problems. So what exactly is his discovery? Maybe it's how to sell Himalayan goji juice to the North American public. And what evidence is there that the juice of this Asian berry has the miraculous healing properties claimed? Here is one line of supporting evidence from Mindell. During the T'ang dynasty (around AD 800), he says, a well had been dug beside a wall near a famous Buddhist temple that was covered with goji vines. Over the years, countless berries had fallen into the well. Those who prayed there had the ruddy complexion of good health, and even at the age of 80 they had no white hair and had lost no teeth, simply because they drank the water from the well. Not convinced? Well, Mindell also tells the tale of Li Qing Yuen who, according to him, represents the best-documented case of longevity. Li Qing Yuen was born in 1678 and lived to the age of 252, marrying fourteen times. How did he manage to do this? He consumed goji berries daily!

According to the numerous websites that sing the glory of goji juice, it took Mindell years of research to perfect his product, which of course is superior to those produced by his imitators. Just what kind of research was this? Did Mindell don a lab coat, like in his promotional pictures, and work in the lab? Did he organize clinical trials? If he did, there is no record of them in the published scientific literature. Did he run case-control studies to see if people consuming the juice were protected from disease? None that I can find. But, of course, such scientific niceties were not necessary, because the promotional sites and pamphlets are careful to state that the product is not intended to treat or cure any disease.

This isn't the first time we've heard of the miraculous properties of some esoteric juice. There's noni juice, mangosteen juice, and, believe it or not, even pickle juice. The claims are usually the same. There is always talk about the abundant concentration of vitamins, the perfect blend of amino acids, the minerals, the antioxidants, and the special ingredients, which in the case of goji are "immune fortifying polysaccharides." And there are references to studies that usually turn out to be experiments in test tubes that show some physiological activity. You can find such results for virtually any fruit or vegetable you can think of. The question is, are there any human clinical trials that show efficacy for the health benefits claimed? A check of the medical literature reveals one Chinese study in which an extract of goji improved the outcome in cancer patients being treated with chemotherapeutic drugs. That hardly qualifies goji juice as a miracle drink. But maybe it's useful in other areas. After all, many goji websites cite an old Chinese proverb that cautions men who are travelling without their wives: "He who travels 1,000 kilometers from home should not eat goji!" Why? Because, as promoters say, goji supports many systems in the body, including ones that control the sex drive. I don't know about that, but there is something else that goji supports very nicely: Earl Mindell.

KOSHER FOOD HYPE

Producers know labels that promote a food as "natural," or "organic," or free of cholesterol, or low in trans fats will often improve sales. These descriptions are now being joined with increasing frequency by the declaration of "kosher" on labels, as marketers try to capitalize on the impression that kosher foods are cleaner and healthier than their counterparts. Indeed, in North America, some five million non-Jews already buy kosher food, and the market increases every time newspapers feature a headline about mad cow disease, *Salmonella* infection in chickens, or shellfish harvested in polluted waters. Surely, consumers believe, when you have to "answer to a higher authority," as claimed in those phenomenally successful ads for Hebrew National hot dogs, you must produce healthier foods? Actually, eating kosher has very little to do with health for the body and more to do with health for the spirit.

It is commonly believed, even by many Jews, that the dietary laws laid down by Moses, as inspired by a "higher authority," were meant to protect people from illness. Before examining this possibility, we need to have some familiarity with the laws of *kasruth*. First of all, let's dispel a myth. It is not a rabbi's blessing that makes a food kosher. A food is kosher when it is prepared according to strict guidelines as first laid down in the Old Testament and elaborated upon by generations of scholars. Although for rigorous followers there are many nuances, the essence of the laws is as follows. Only mammals that chew their cud and have a split hoof can be eaten. So cows and sheep are in, pigs and rabbits are out. Domesticated fowl such as chickens and ducks are fine, as are fish with fins and scales, but insects are forbidden. Meat cannot be eaten with dairy products, but eggs, fruits, vegetables, and grains can be consumed with either meat or dairy. Utensils that have come into contact with meat must not be used for dairy and vice versa. Animals have to be killed by hand with a very sharp knife, all blood must be drained, and the carcass has to be inspected for disease.

The argument that kosher foods are healthier usually focuses on pigs. These animals wallow in dirt, sometimes eat excrement, and can harbor the parasites that cause trichinosis. Accordingly, some claim, God, through Moses, forbade their consumption. Actually, cattle can transmit an even larger variety of nasty organisms, such as tapeworms, *E. coli* bacteria, and anthrax. Chickens will peck at excreta and commonly are infected with *Salmonella* or *Campylobacter*, which can cause human disease. There is no reason to believe that cooked pork is any more risky than other meats. But it was not easy to raise pigs in the open desert. They require shade and, unlike cattle or sheep, cannot live on dry grass and coarse shrubs. Pigs need some seeds or tubers, foods that can also be eaten by humans. Essentially, pigs would not have been a wise investment for farmers during biblical times. As far as combining meat with dairy products is concerned, there is no scientific evidence of digestive problems. Furthermore, there certainly is no health issue involved when a steak is eaten off a plate that has previously been used for cottage cheese. Why, then, did scholars, even the great Moses Maimonides, suggest that "pork has a bad and damaging effect on the body?" Surely, even in the twelfth century they realized that "kosher" animals were as likely to carry disease as others. In all likelihood, such declarations were made with the deeply held belief that Moses and his "advisor" didn't give commandments for arbitrary reasons.

Indeed, the commandments had a purpose, but that purpose had to do with religious discipline, not health. The laws of *kasruth* were meant to ensure that even commonplace activities such as eating had a spiritual connection. Following them emphasized the constant presence of God and the need to follow his commandments at all times.

Does this mean that the millions who buy kosher foods for non-spiritual reasons are wasting their money? Not necessarily. For example, if someone is allergic to shellfish, they can safely eat any kosher food. A kosher logo that carries the symbol "D" signals the presence of dairy products, so the absence of the D means that such foods are fine for anyone who needs to avoid dairy due to lactose intolerance or allergies. Kosher chickens may have a lower bacterial count because the salting process kills many microbes, but proper cooking renders chickens safe in any case. Kosher poultry does tend to be fresher and often has a better taste. An interesting issue has recently arisen in connection with mad cow disease. Conventional slaughter involves striking the animals on the head, a process that some claim can scatter brain tissue, along with the prions that cause mad cow disease, throughout the bloodstream. This cannot happen when a kosher butcher, or *shohet*, slits an animal's throat with one quick slice.

Kosher does not mean that the animals have been raised without the use of hormones or antibiotics, nor does it signify the absence of additives. It does mean that certain additives, such as the red dye carmine (derived from a species of insect), are not used. It also means that a product like Coke, labelled as kosher, cannot contain any substance derived from nonkosher animals. So the glycerin, one of Coke's flavor components, must come from a vegetable source. Most assuredly though, "kosher" does not mean nutritionally superior. Hebrew National may have to answer to a higher authority about how its hot dogs are prepared, but those dogs are still loaded with fat and salt. As with any other hot dog, if you eat too many, you may get to discuss the fine points of *kasruth* with that higher authority sooner rather than later.

THE QUESTIONABLE HEALTH
PROPERTIES OF DHEA

"This could spell trouble!" That undoubtedly was the sentiment reverberating through the offices of the Council for Responsible Nutrition after publication of a study about dehydroepiandrosterone (DHEA) and aging in the *New England Journal of Medicine* in 2006. The council is a lobby group sponsored by the dietary supplement industry, and DHEA has been one of the industry's golden boys, raking in millions of dollars in sales annually. No surprise here, given that promoters have successfully painted an image of DHEA as a "fountain of youth" hormone.

Interest in DHEA was originally generated with the discovery that production of this substance in the body peaks in our twenties and then tails off. By the time we reach our seventies, only about a fifth as much DHEA circulates in our bodies as during our youth. Is it possible then that slowing this decline can have an anti-aging effect? Certainly a reasonable question to ask, especially since DHEA is known to be involved in the production of both the male and female sex hormones, and these certainly have important functions in the body. DHEA is made from cholesterol in the adrenal glands and serves as a precursor to estrogens and testosterone. Although DHEA itself is often called a hormone, it does not fit the definition. Hormones are chemical messengers that trigger some sort of physiological activity in the body at a location remote from where they are synthesized. This has never been shown for DHEA. But that does not preclude its involvement in the aging process. Early hope was that animal experiments would clarify the situation.

Initial trials with rodents were encouraging. Actually, the effects of DHEA seemed almost miraculous. Rats and mice given the supplement showed a decline in obesity, an improvement in immune function, and a reduced risk of heart disease and cancer. But the relevance to humans of these effects was questionable from the beginning, because rodents hardly produce any DHEA at all, meaning that the doses given were huge relative to the amount of naturally circulating substance. Still, the rodent data were interesting enough to stimulate human research. There was excitement when Dr. Elizabeth Barrett-Connor at the University of California found that men with high levels of DHEA were less likely to die of heart disease. And supplement manufacturers were absolutely thrilled when Dr. Samuel Yen, also at the University of California, carried out a placebo-controlled trial over three months in eight men and eight women between the ages of fifty and sixty-five, and found some positive changes in immune function and an improved sense of "well-being" in the DHEA group.

This was enough to crank up the advertising machinery, and soon DHEA supplements appeared in health food stores despite protests both from Barrett-Connor and Yen that their work was preliminary, and that there were too many unknowns about DHEA to recommend its use. Dr. Richard Weindruch of the Medical College of Virginia, whose studies on the longevity of mice were prominently quoted in the DHEA promotional literature, joined the fray and explained that his work was taken out of context and that his mice in fact did not live longer. Much of the hype centered on DHEA's supposed ability to cause weight loss. This did not sit well with the US Food and Drug Administration because such a claim would make DHEA an unapproved new drug. Warnings went out to remove the substance from the market. But DHEA re-emerged with more claims than ever after the passing of the Dietary Supplement Health and Education Act in 1994, which curiously allowed DHEA to be classified not as a drug, but as a food supplement. Why? Because it occurs naturally in meat; therefore, it is a "food." Canada has a more sensible approach and does not allow DHEA to be sold, maintaining, correctly, that the claims on its behalf are not nutritional but pharmacological. No doubt, though, many Canadians, seduced by the advertising for the rejuvenating "superhormone" purchase it outside the country by mail order.

Ads for DHEA are generally cleverly written and do refer to studies, but they do not give the complete picture. No mention is made of the short duration and few subjects in the trials, or of the potential side effects of altering hormone levels in the body. Now perhaps the study by Mayo Clinic researchers published in the *New England Journal of Medicine* will put the brakes on the runaway DHEA bandwagon. This two-year, placebo-controlled trial is the longest and best study ever carried out on the supplement. True, it did not examine every effect that has ever been claimed on behalf of DHEA. Libido was not investigated, and other possibilities, such as benefits in diseases such as lupus, will have to wait for other studies.

Unlike previous trials, the Mayo Clinic investigation didn't involve only a handful of subjects; eighty-seven elderly men and fifty-seven elderly women were enrolled. The results? A seventy-five-milligram daily dose of DHEA increased blood levels of the substance as expected, but had no effect on oxygen consumption, insulin sensitivity, muscle strength, or body composition, all accepted markers for aging. A slight effect on bone mineral density was noted, but according to the researchers it was minimal and inconsistent. In any case, this minimal effect pales in comparison to what can be achieved by other medications.

These results were not what the dietary supplement industry had hoped for. So the spinmeisters at the Council for Responsible Nutrition went to work. "This is the longest duration human supplementation trial confirming the safety of relatively high-dose DHEA in both men and women," a press release triumphantly proclaimed. Basically the message was, you can keep taking DHEA because an excellent study has shown it to be safe! No mention of the fact that the "excellent study" found the substance to be useless. Unfortunately, such spin-doctoring is commonplace these days, usually on both sides of a scientific issue. And is enough to make your head spin.

ALKALINE NONSENSE

If you want to protect yourself against cancer, just eat right. We've certainly heard that advice before. But what does eating right mean? According to some alternative practitioners, all we have to do is eat an "alkaline" diet to ensure that our body is maintained in an "alkaline" instead of an "acid" state. It sounds so seductively simple. When a cell becomes cancerous, proponents of this theory claim, it reduces its use of oxygen and cranks up its production of acids. These conditions then allow cancer cells to multiply quickly. What can we do to prevent this from happening? Ensure that cells get an adequate supply of oxygen, and that the acids produced are neutralized! How? By introducing sources of oxygen such as hydrogen peroxide or ozone into the body, and consuming "alkaline" foods. If cancer has already taken a foothold, then it may be necessary to dose up on cesium, the "most alkaline nutritional mineral." So simple—and so wrong!

As so often happens, promoters of nonsensical therapies seize a few filaments of scientific fact and weave these into a tangled web that ensnares the desperate and the scientifically confused. In this case, it starts with the work of German physician Otto Warburg, who received the 1931 Nobel Prize in medicine for his work on cellular metabolism. Warburg showed that the growth of malignant cells requires markedly smaller amounts of oxygen than that of normal cells and that their metabolism follows an anaerobic (that is, not requiring oxygen) pathway leading to the production of lactic acid.

This notion lay dormant until the 1980s when Dr. Keith Brewer, a physicist with no medical training, used it to support his perplexing theory of how potassium and calcium control the transport of glucose and oxygen into cells, and how irritation of the cell's membrane interferes with this transport system. The result, Brewer maintained, is the "Warburg effect," which increases a cell's acidity (and lowers its pH), reduces its oxygen supply, and causes changes in DNA characteristic of cancer. He then went on to claim that cesium's chemical similarity to potassium allows it to be readily taken up by cells but, unlike potassium, it does not transport glucose into cells while allowing oxygen to enter. As a result, cancer cells are enriched in oxygen, deprived of glucose, form less lactic acid, become more alkaline and, as a consequence, die. Sounds good, but Brewer got the "Warburg effect" all wrong. Cancer cells do shift to a mode of metabolism that doesn't use oxygen, but this happens even in the presence of oxygen.

Brewer went on to buttress his argument by claiming that cancer is almost unknown among the Hopi Indians of Arizona, the highland Indians of Peru, and the Hunza of North Pakistan. Why? Because due to cesium in the local soil, they have a "high pH" diet. Whether these people actually do have a lower cancer rate is questionable, and even if this were the case, it could not be ascribed to cesium in the diet without further investigation. But then to take the cake (undoubtedly cesium enriched), Brewer in 1984 published a paper with the following claim: "Tests have been carried out on over thirty humans and in each case the tumor masses disappeared. Also, all pains and effects associated with cancer disappeared within twelve to thirty-six hours; the more chemotherapy and morphine the patient had taken, the longer the withdrawal period." Not only had he discovered the cancer cure that had eluded the thousands of PhDs and MDs working in cancer research around the world, but he had also shown that chemotherapy was actually harmful.

Where were these miraculously cured patients and who had treated them? Brewer refers to Dr. Hellfried Sartori (a.k.a. Professor

Abdul-Haqq Sartori), who had accomplished this incredible feat in the Washington DC area. This is the same Dr. Sartori who in July of 2006 was arrested in Thailand and charged with fraud and practicing medicine without a license. He was charging desperate patients up to $50,000 for "cancer cures" that included cesium chloride injections. The good doctor, who routinely promised that he could cure his patients of any disease, has a rather illustrious history. Known as the notorious "Dr. Ozone" in the United States, he served five years in prison in Virginia and nine months in New York for defrauding patients with unapproved therapies such as cesium chloride injections, coffee enemas, and ozone flushes. Needless to say, there are no records of the patients whom, according to Brewer, Sartori cured of cancer. Australian police are now looking into the deaths of six people who died after intravenous injections of cesium chloride at clinics following Sartori's protocol.

Raising a cell's pH with cesium chloride makes no sense, scientifically, but this is not what rules out the treatment's possible effectiveness—it's the lack of evidence that does. There are no controlled trials showing cancer being cured with ozone or cesium—but there is evidence that cesium chloride can cause cardiac arrhythmia and death. Granted, it is unlikely that this can happen from the oral doses being promoted by numerous alternative practitioners aimed at raising the body's pH, but the idea that cesium chloride can neutralize acids in cells is sheer nonsense.

Yes, cesium is an "alkali" metal. Dropping a piece of cesium metal into water does indeed produce an alkaline solution. But cesium chloride is not the same as cesium metal; the former is a neutral salt. In any case, the blood's pH cannot be altered by cesium chloride ingestion, or indeed with the ingestion of any food. Human blood chemistry is a marvelously buffered solution, meaning that it resists any change in acidity. It doesn't matter what we eat or drink, our blood contains substances that can act as acids or bases to maintain our blood pH at 7.4. The only body fluid that responds to diet in terms of pH is urine. Breads, cereals, eggs, fish, meat, and poultry can make the urine more acidic while most, but not all, fruits and vegetables make the urine more alkaline. A diet high in fruits and vegetables and low in meat can indeed reduce the risk of cancer, but this has absolutely nothing to do with changing the pH of cancer cells. The idea of an "alkaline" diet to prevent or treat cancer may sound seductively simple, but in reality it is just simpleminded.

LOSING WEIGHT WITH GREEN TEA?

Soft drink producers are in a quandary. Their product is coming under increasing nutritional scrutiny and it is not faring well. Schools are eliminating the sales of soft drinks, and the public is becoming increasingly wary of consuming sugar-laden beverages with "empty calories." Replacing sugar with artificial sweeteners does not seem to be the answer to the marketing woes, mainly because of the common (generally unjustified) perception that these substances are mired in unresolved safety issues. So if empty calories or zero calories don't boost sales, how about "negative calories?" A beverage that causes more calories to be "burned" than it supplies certainly sounds attractive. And the Coca-Cola Company claims that it has come up with just such a product in Enviga, its new green-tea-based drink.

According to Dr. Rhona Applebaum, chief scientist for Coke, "Enviga increases calorie burning and represents the perfect partnership of science and nature." Let's take a look at this "perfect partnership." First, calories cannot be "burned"; they are not things, they are a unit of measure. Simply stated, a food calorie is the amount of heat needed to raise the temperature of a kilogram of water by one degree Celsius. Where then does the expression "burn calories" originate? When a substance burns, it releases heat. If a piece of pie is said to contain, say, 300 calories, then combusting it in a closed chamber, called a calorimeter, will produce enough energy to heat 300 kilograms of water by one degree.

Our body can also "burn" that piece of cake, meaning that 300 calories worth of energy is released as a series of chemical reactions decompose, or metabolize, the cake's fats, carbohydrates, and proteins. The products of these reactions are eventually exhaled in our breath or excreted in the urine and feces, while the energy produced is used to maintain our body temperature and supply the power needed for the proper functioning of our organs and muscles. If we do not "spend" all the calories that are potentially available, there is no need for the body to completely "burn" the food components, and the remnants are stored. Weight gain ensues. If we engage in activity, the stored supplies can be called upon to undergo the reactions needed to produce the required energy, and weight is lost. Obviously then, to lose weight, more calories must be expended than are provided by the ingested food.

Three servings of Enviga (one serving is 330 milliliters) contain only fifteen calories, but according to Coca-Cola the drink stimulates the body's metabolism to produce an extra 60 to 100 calories per day. These calories, given off in the form of heat, are produced when stored nutrients are converted to substances that are released from the body. The implication is that drinking three servings of Enviga a day will lead to weight loss, although the company is careful not to make that claim. Of course it does hope that the prospect of easy weight loss will make the product fly off the shelves.

Time now to look at the science behind the hype. It all started back in 1999 when researchers at the University of Geneva made an interesting observation about the inhibition of an enzyme, catechol O-methyltransferase, by catechins, compounds found in green tea. This enzyme degrades the neurotransmitter norepinephrine, which stimulates fat oxidation and heat production (thermogenesis). If norepinephrine breakdown is curbed, the thinking went, thermogenesis should be increased, potentially leading to weight loss. This reasoning seemed to mesh with the observation that Asians are great green-tea consumers and are rarely overweight. Why not then try to give volunteers green-tea catechins in a dosage roughly comparable to what Asians consume, and then monitor their energy expenditure?

The standard technique is to place subjects in a respiratory chamber, which is a completely sealed room that allows the incoming and outgoing air to be monitored for carbon dioxide and oxygen levels. The "combustion" of nutrients in the body requires oxygen and produces carbon dioxide and energy (calories). Since the amount of energy produced relative to oxygen uptake and carbon dioxide released is known, total energy expenditure over a twenty-four-hour period can be determined. When such an experiment was carried out using ten male volunteers who were given capsules every day containing 375 milligrams green-tea catechins, their energy expenditure was increased by about eighty calories. Not very impressive, but still, scientifically meaningful and enough to spur other studies. And it is one of these studies that Coca-Cola uses to promote Enviga. Fifteen men and sixteen women consumed a prototype beverage three times a day containing a total of 540 milligrams catechins and 300 milligrams caffeine, which also is known to boost metabolism. Energy expenditure went up by about 100 calories a day without any change in heart rate or blood pressure, which was comforting. Since the testing period was only three days, no weight loss was noted. This study was quite small and has not yet been published in the scientific literature, which is sort of curious given the extent of the Enviga marketing campaign.

A double-blind Japanese study in 2005 did show some weight loss with green-tea extract. Half of thirty-eight employees (all male) of the Kao Corporation drank a green-tea beverage laced with 690 milligrams of catechins every day with dinner while the others had the same tea laced with only 22 milligrams of catechins. All the men were put on a diet with 10 percent fewer calories than needed to maintain their weight. Over three months, the catechin consumers lost 1.1 kilograms (2.4 pounds) more than the men who drank conventional tea. Interesting. And guess what the Kao Corporation makes? Catechin-fortified green tea! In Japan the company has even been allowed to make the label claim: "Due to its high content of tea catechins, this green tea is suitable for people concerned about body fat." But if you are really concerned about body fat, eat less and exercise more. Is it worthwhile to quench your thirst with Enviga after your workout? Can a beverage with only ninety milligrams of catechins per serving result in any appreciable weight loss? Fat chance.

THE MYTH OF "DETOX"

No wheat, no meat, no dairy, no alcohol, no caffeine, no sugar, no salt, no processed foods. Lots of fruits and vegetables, wheat-free pasta, brown rice, nuts, seeds, beans, lentils, tofu, lemon juice, and liters of water. What do you call such a diet? "Detox," say the food faddists. "Bizarre," say serious nutrition researchers. Detox advocates claim that our modern lifestyle floods the body with toxins, although their definition of this term is somewhat confused. It seems what they mostly have in mind are pesticide residues, food additives (despite the stringent regulations that govern these substances), and environmental pollutants, such as PCBs, dioxins, plasticizers, and mercury. But sugar, salt, meat, and dairy also get tossed into the mix as toxic substances. All these toxins, the detoxers claim, build up in our tissues and conspire to bring on weight gain, headaches, bloating, fatigue, lowered immunity, and dull-looking skin. We are doomed, they say, unless we periodically flush these toxins from our bodies. And the way to do that is through a detox diet.

But where is the evidence? Has anyone carried out studies to show that "toxins" appear in urine or feces or sweat after a detox diet? I can't find any such data. The fact is that our bodies are engaged in detoxification all the time. Our liver and kidneys are very adept at removing undesirable intruders, be they synthetic or natural. Is it perhaps possible that a detox diet can increase the efficiency of these organs? After all, there are people who claim they feel better after such a regimen. So, sniffing a potentially hot story, the BBC decided to put the detox diet to a test. Producers of *The Truth About Food* tracked down ten women aged nineteen to thirty-three who had been partying at a rock festival and were obvious candidates for a detoxification experiment.

Five of the women were put on a classic detox diet while the others followed a regular, healthy diet. All the subjects then sacrificed some of their body fluids for the sake of scientific research. Creatine levels were measured in the urine to monitor kidney function, and blood was tested for liver enzymes to determine the health status of that organ. Blood was also tested for vitamins C and E, indicative of antioxidant potential, as well as for aluminum, which is often targeted as a significant toxin by detox proponents. No significant differences were noted between the groups. There was no apparent detoxification. How then is it that some people claim that they feel rejuvenated after a detox cleanse? Caffeine and alcohol can cause headaches, so eliminating these may be of help. Less food consumed can relieve bloating, and paradoxically, near-starvation can trigger a boost in energy and even feelings of euphoria. This is probably an evolutionary vestige from the times when hungry people had to muster up a last bit of energy in an attempt to locate food.

Even if detox diets do result in improved feelings of well-being, their concept is flawed. The message is that our body will forgive our dietary sins if we periodically undergo a cleanse. That's not what sound nutrition is all about. Focus should be on eating in a healthy fashion all the time, not on making some dramatic alteration when a problem arises. But that idea doesn't sell nearly as well as claims of miraculously restored health by a short-term change in diet. The dramatic tale told by anesthesiologist Anthony Sattilaro in his bestseller *Recalled by Life* is a case in point.

Dr. Sattilaro was diagnosed with widespread cancer back in the late 1970s. As luck would have it, he picked up a hitchhiker who had just graduated from a natural cooking school. The young graduate told the doctor that he didn't have to die, that cancer wasn't that hard to cure. And thus began Sattilaro's plunge into the world of macrobiotics. Desperate people will do desperate things. So out went meat, dairy products, fruit, oil, and eggs; in came brown rice, boiled vegetables, black seaweed, miso soup, and pickled plums.

Almost instantly, pain that had been controlled with heavy-duty drugs disappeared, and within three years, so, apparently, did the cancer. *Recalled by Life* became a bestseller and launched numerous cancer patients down the hopeful path of macrobiotics. Needless to say, patients who followed in Sattilaro's footsteps but had no reversal of their fortunes did not end up writing books about their experiences. Alas, Sattilaro's cancer returned, and this time no diet was able to save him. Did the "detoxing" macrobiotic diet cause the original turnaround? Who knows? Sattilaro also had surgery to remove his testes, prostate, and a rib, and received estrogen therapy.

Dr. Sattilaro was not the first, nor the last, claiming to have found the secret of restoring health by detoxifying the body. In the 1950s, Adolphus Hohensee urged people to insert a clove of garlic into their rectum every evening to rid the body of toxins, and suggested that the scent of garlic on the morning breath was proof that the detoxifying chemicals had worked their way through the body. In the seventies, Durk Pearson and Sandy Shaw, "leading independent experts in anti-aging research and brain biochemistry," in their bestseller *Life Extension*, urged us to consume some thirty dietary supplements a day. David Steinman came along in the 1980s with his *Diet for a Poisoned Planet*, recommending megadoses of niacin to counter the alleged effects of pesticides and industrial chemicals in our food.

The eighties also brought us Harvey and Marilyn Diamond's *Fit for Life*, which claimed that not eating starches and proteins together was an important step towards detoxification. There is no scientific evidence for this. The first decade of this century introduced us to naturopath Peter D'Adamo's ideas about *Eating Right for Your Blood Type*. Women with type A blood and a history of breast cancer can benefit from eating snails, he suggests. Again, there is nothing in the published scientific literature to back up this claim. Alex Jamieson in her *Great American Detox Diet* (she's the girlfriend who restored

Morgan Spurlock to health after he supersized himself by eating exclusively at McDonald's for a month) reminds us of art class where we made papier-mâché with flour and water. This is just like the goo that forms in our body if we eat white bread, she claims, without providing any evidence. No white bread for Alex, but lots of sea veggies, which can clean out the body. Yummy. I can only hope that the next detox scheme that emerges is more palatable both to the mind and the body.

WHOM TO BELIEVE?

"The worst kind of ignorance is the things we know for sure that just ain't so." I'm not certain just what Mark Twain had in mind back in the 1800s, but today his clever quip could apply to some of the nutritional "information" being spread around. How can anyone distinguish the sense from all the nonsense out there? Considering that most scientific issues are not white or black but various shades of grey, there can be no simple answer to this question. Nobody has a monopoly on truth. Still, our best bet is to formulate our opinions based on consensus derived from the peer-reviewed scientific literature as published in reputable journals. Unfortunately, when it comes to conveying information to the public, scientists tend to quietly—and often boringly—recite data, while activists bellow from atop their soapboxes. But repetition of dogma and emotional outbursts should not be confused with science. It may be helpful to examine a specific example.

Artificial sweeteners such as aspartame and sucralose, as we have seen, are controversial. Their opponents would have us believe that they should be avoided. Their defenders maintain that when used as directed, they can be helpful for diabetics, as well as for people looking to reduce their caloric intake.

So, who is who in this battle? On one side, we have the US Food and Drug Administration, Health Canada, and the regulatory agencies of some eighty countries around the world. These are staffed by a selection of PhDs and MDs trained in chemistry, biology, toxicology, physiology, and epidemiology. On the other side we have an eclectic mélange of personalities. Here are some of the people who dominate the antisweetener crusade: Dr. Janet Starr Hull, Dr. Betty Martini, Dr. James Bowen, and Dr. Joseph Mercola. Let's meet them.

Dr. Hull received her doctorate in nutrition from Clayton College of Natural Health, a nonaccredited correspondence school. It offers courses in detoxification and healing, iridology, homeopathy, and human energy fields. Conveniently, the college even sells healing products online. Not only can students, or indeed anyone else, purchase a variety of homeopathics and herbal supplements, they can even load up on supplements for their companion animals. The college does offer a course on basic chemistry (and she likely took some courses in chemistry while pursuing her degree in environmental science), but Dr. Hull shows no evidence of its subject matter when she makes statements such as "Splenda is one-fourth sugar, three-fourths chemical" and that "chlorine found in nature is different from chlorine that has been manmade and adulterated." Hull also explains that in order for volatile chlorine to be "locked in," manufacturers of sucralose rely on acetone, benzene, formaldehyde, and methanol, all of which are "used" in gasoline and petroleum. What a cacophony of nonsense!

Hull is trying to imply that sucralose is toxic because it contains the "deadly chemical" chlorine. Yes, sucralose does contain chlorine; in fact, each molecule has three chlorine atoms. But these are bonded to the framework of a sugar molecule and have nothing whatsoever to do with chlorine gas. So Hull is plainly wrong when she says that to understand the ill health "caused" by sucralose you "must look for chlorine poisoning symptoms." No chlorine gas is released from sucralose; in fact about 85 percent of a dose is completely unabsorbed by the body. The rest is broken down to simpler compounds, but there is no dechlorination and no chlorine is retained in the body.

Dr. Betty Martini, whose mission is to rid the world of nasty substances such as aspartame and sucralose, also addresses the chlorine issue. It seems she knows more chemistry than the manufacturer of sucralose, as she opines in a letter to the company: "If you don't understand the dangers of chlorine, then you need to step down as a manufacturer or start calling your product DDT-Lite. Do you think the consumer public is so stupid they don't understand that sucralose is a chlorocarbon poison?" Martini even offers to send her documentation of adverse effects to the company executives and researchers in Braille, since they must obviously be blind, unable to read the evidence about the adverse effects of chlorinated substances like DDT. Yes, DDT is a chlorinated compound, but this has *nothing whatsoever* to do with sucralose. Toxicity is determined by the exact three-dimensional structure of a molecule, not by what atoms it is composed of.

Dr. Martini attempts to back up her views with references to the work of others. Repeatedly, she brings up the work of Dr. James Bowen, described as a noted "physician, researcher and biochemist." There is no record of this researcher having published anything in the scientific literature, but he has "researched sweeteners for twenty years after discovering that he developed Lou Gehrig's disease after being poisoned with aspartame." He regards chlorine as "nature's Doberman attack dog, a ferocious atomic element employed as a biocide, as a World War I poison gas and a reagent to make hydrochloric acid." None of this has anything to do with sucralose, but there is more to Dr. Bowen than chemical ignorance. It seems that substances like aspartame and sucralose are being unleashed on the American public to effect "mind control." Who is behind this? According to Bowen, Zionists. "They see it as their patriotic duty to Zionism and Israel to see to it that we succumb to aspartame! Masons and Satanists have likewise done everything they could to destroy me and my ministry." Bowen goes on to rant that "aspartame's marketing by [Donald] Rumsfeld (once president of the company that sold the sweetener) was an organized crime, protected by Zionists, Mossad, B'nai B'rith, Masonry and all other satanic organizations." He also maintains that the sinking of the *Titanic* was a plot to kill influential Christians and that the Twin Towers were brought down by explosives in a clever plot engineered by Satanists such as President George W. Bush.

Bowen is also referenced as an authority on toxicity by osteopath Dr. Joseph Mercola, who maintains a popular health website and sells a variety of supplements. In all fairness, I doubt that Mercola is aware of Bowen's toxic anti-Semitism, and hopefully will expunge any reference to this disturbed individual after doing a little checking. Mercola bases his antisucralose arguments on undocumented anecdotal accounts, the tired argument that sucralose, like PCBs, contains chlorine, and that the studies used to prove the sweetener's safety were inadequate. The message is that all chlorinated compounds are bad. (Wonder if he's ever heard of vancomycin, a wondrous antibiotic that contains chlorine.) But then again, osteopathy may not be the most appropriate preparation for an analysis of complex scientific studies. Or, for that matter, of nutritional concepts. Recently, Mercola has received two letters from the FDA warning him to stop making illegal claims about his supplements' ability to cure or mitigate disease. In response, he changed the wording of the claims to abide by the letter of the law.

I am no great fan of artificial sweeteners, mainly because they take the focus away from promoting an overall healthy lifestyle. They are not the answer to our obesity problem. And in rare cases, they can, like any substance, cause adverse health effects. But when it comes to evaluating their overall risk-benefit ratio, whom would you rather trust: the peer-reviewed scientific literature, or the various ramblings of Drs. Hull, Martini, Bowen, and Mercola?

CONCLUSION:
IS THERE A SOLUTION TO THE CONFUSION?

Phew! That was a lot to digest, wasn't it? And where do all these ruminations on food leave us? It seems that with the plethora of scientific studies being carried out these days, evidence can be found to support virtually any point of view. But one must always be wary about putting too much emphasis on single studies; they rarely produce giant leaps in science. The unromantic truth is that science plods along via a series of small steps, hoping eventually to bring about a consensus of expert opinion. In the case of nutrition, that has more or less happened. And the consensus is quite clear. Eat lots of fruits, berries, and vegetables, aiming for eight to ten servings a day. Wash them well and don't fret about whether they have been organically or conventionally grown. Look for variety: the more colorful, the better. Eat fish a couple of times a week, mindful of the fact that women of child-bearing age and young children need to limit intake of species such as swordfish and fresh or frozen tuna, which are known to be high in mercury. Red meat should be an occasional treat; poultry is preferable. In either case, these should cover a small portion of the plate, the rest being piled with vegetables, brown rice, or whole-grain pasta. Start most days with oats, flax, and berries. No need to be horrified by eggs: even five a week are unlikely to have an effect on blood cholesterol. Minimize processed foods, particularly those that are high in salt and hydrogenated fat. Low-fat dairy products are a great source of calcium and should be included in the diet. Soft drinks have no redeeming nutritional value.

Green tea is a great beverage, though there are few problems with coffee, if consumed in moderation. Nuts are excellent snacks. Use canola or olive oil, but avoid frequent frying or barbecuing. Dark chocolate is a better dessert than chocolate cake. One alcoholic beverage a day is fine. It goes without saying that overall calorie intake has to be balanced with energy expenditure. And remember that there are no "miracle" foods or beverages out there.

Not that complicated, is it? But other facts come into play. Most people's taste buds would vote for a hamburger over a veggie burger, french fries over lentils, brie over low-fat cottage cheese, and an apple danish over an apple. And if occasionally you feel like indulging your taste buds, well, go ahead. After all, as I've said before, there is more to life than worrying about every morsel of food we put into our mouths. What matters is the overall diet. It is possible to eat apples every day and still have a nutritionally nightmarish diet, just as it is possible to eat the occasional doughnut while maintaining a good diet.

Exactly what constitutes a good diet is constantly being fine tuned. While the guidelines offered above are based on solid science and are unlikely to be dramatically changed with future research, refinement is certainly possible. We have recently learned, for example, that while cinnamon may be of some value in helping type-2 diabetics control their blood sugar, it doesn't work for type-1. On the other hand, apples may be even better for us than we thought. A recent study showed that women who eat apples during pregnancy may protect their children from developing asthma later in life. We've also heard about the possibility of developing a genetically engineered tomato that is particularly high in folic acid, which would make recommending its consumption reasonable.

When you carefully scrutinize the scientific studies that are being rolled out almost on a daily basis, most amount to no more than tinkering with the basic nutritional principles we have tried to lay down: eat mostly foods based on vegetables, fruits, whole grains, and low-fat dairy products, and don't overeat.

That's why tomorrow I'll have my oatmeal in the morning, sprinkled with ground flaxseeds, topped with berries, and washed down with orange juice. I'll have a tomato, lettuce, and cheese sandwich on whole-grain bread for lunch with some hummus, a banana, and a pear. (Were I not allergic to fish, I'd probably have some canned tuna or salmon.) Snacks? Unsalted nuts, carrot sticks, and live-culture yogurt. Beverages? Water, coffee, or tea. For supper, I'm thinking of bean and barley soup, spinach salad, chicken paprikas, along with my newly developed broccoli, tomato, and brown rice casserole. Dessert? Strawberries and grapes. Maybe dipped in dark chocolate. And then I'll go to sleep and dream of a smoked meat sandwich, french fries, and a dill pickle. (Occasionally, I'll even make this dream come true.) Oh yes, something I almost forgot, something that I eat every day: an apple.

INDEX